Praise for Peter Pri...

"In this elegantly succinct and well-research... nineteenth century laboratory of the Morav... rooms, town ... and universities whe... conglomerates like Calgene and Monsanto.... *Food, Inc.* proves beyond doubt that you really know more when you are able to admit how little it is that you actually know."

—*Los Angeles Times*

"Indeed, one of the most valuable things about this book is the way Pringle excoriates both sides in the burgeoning Frankenfood debate, both the multinational corporations intent mainly on promoting their products and the environmental activists bent on scaring consumers.... A well-researched tome ... worth reading if you care about what goes on your plate."

—*Austin American-Statesman*

"Far more informative and better-written [than Philip Yam's *The Pathological Protein* and Andrew Rowell's *Don't Worry (It's Safe to Eat)*] is Peter Pringle's new book on GM crops.... Mr. Pringle carefully dissects the anatomy of resistance to GM food in Britain, as well as recent scares in America over contaminated taco shells and disabled butterflies. His book also clearly lays out the environmental and economic uncertainties associated with GM crops and the risks of allowing GM technology to propagate the 'plague of sameness' which has come to characterize modern, industrialized agriculture. The book offers an interesting view of the origins and current deficiencies in America's laissez-faire approach to regulation.... Despite his tough stance, Mr. Pringle takes a realistic view of what GM technology can do, particularly in addressing the agricultural challenges of the developing world."

—*The Economist*

"*Food, Inc.* is the veteran journalist Peter Pringle's rendering of the wars over these scientific wonders. This he does with the eye of a curious traveler; he's a man who sees this colorful cast of academic climbers, corporate suits, English seed-huggers and Indian firebrands as a giant circus of savants—folks so intent on their own little world of passionate specifics that time and time again they fail to see the larger and more important picture.... Pringle also ministers a well-earned cuffing to the many activist academics whose narcissistic politics helped cloud public understanding of genetic modification.... Delivering a slim volume on a big subject, Pringle has done a fine job, honing the basic scientific issues to their essence in workmanlike fashion and then exploding the controversies at their core.... Anyone who picks ... Monsanto and

- 8 MAR 2006

their tormentors on the picket lines—will find *Food, Inc.* to be a feast of honest reporting and serious thought. It's about time."

—*The New York Times Book Review*

"In *Food, Inc.*, author Peter Pringle reviews an array of genetically modified food case studies—stories that bring together science, politics, business, and biotechnology ethics—in an attempt to shed light on all sides of the debate. . . . Pringle strips connotation-laden terms such as 'gene modification' and 'genetic engineering' down to their biological basics, teaching the reader how these technologies actually work. . . . Pringle goes to great lengths to give a fair hearing to both sides . . . for those who value scientific rigor, who question hype, and who understand that life always carries risks."

—*The Boston Globe*

"So today's corn comes from pampered plants that couldn't survive were it not for the generous helpings of chemicals, water and, in some cases, human manipulation. And the health and environmental soundness of these practices is largely determined by the companies themselves. *Food, Inc.* makes you think about these things. Unlike many books on controversial topics, *Food, Inc.* doesn't impose a heavy-handed point of view. . . . This book contains the kind of detail interested consumers are hungry for. . . . So go ahead and dip that chip into some salsa. Just be sure to read the chapter on tomatoes."

—*The Dallas Morning News*

"Pringle fills in some of the blanks concerning genetically modified food by answering basic questions, demystifying language, explaining the science, and making sense of scare tactics and propaganda."

—*The Futurist*

"Peter Pringle takes on a formidable challenge. . . . *Food, Inc.* presents lively and colorful accounts of events on the biotechnology battlefield."

—*The Washington Post*

"The genetic modification of plants is a controversy ripe with ideological complexity and scientific disputation. Peter Pringle's little book is a good place to enter that thicket. . . . Pringle attempts to consider this argument as calmly and evenhandedly as possible."

—*St. Louis Post-Dispatch*

Also by Peter Pringle

Cornered: Big Tobacco at the Bar of Justice

S. I. O. P.: The Secret U.S. Plan for Nuclear War
(WITH WILLIAM ARKIN)

The Nuclear Barons (WITH JAMES SPIGELMAN)

Those Are Real Bullets: Bloody Sunday, Derry, 1972
(WITH PHILIP JACOBSON)

FOOD, INC.

Mendel to Monsanto—
The Promises and Perils of the
Biotech Harvest

———

Peter Pringle

SIMON & SCHUSTER PAPERBACKS

New York   London   Toronto   Sydney

BARNSLEY LIBRARY SERVICE

10015403	
Bertrams	25.01.06
133.3092	£8.99

SIMON & SCHUSTER PAPERBACKS
Rockefeller Center
1230 Avenue of the Americas
New York, NY 10020

Copyright © 2003 by Peter Pringle
All rights reserved,
including the right of reproduction
in whole or in part in any form.

First Simon & Schuster Paperbacks edition 2005

SIMON & SCHUSTER PAPERBACKS and colophon are registered trademarks
of Simon & Schuster, Inc.

For information regarding special discounts for bulk purchases,
please contact Simon & Schuster Special Sales at
1-800-456-6798 or business@simonandschuster.com

Manufactured in the United States of America

1   3   5   7   9   10   8   6   4   2

Library of Congress Cataloging-in-Publication Data
Pringle, Peter.
Food, inc. : Mendel to Monsanto—the promises and perils of
the biotech harvest / Peter Pringle.
p.   cm.
Includes index.
1. Agricultural biotechnology.   2. Genetically modifed foods.
3. Food—Biotechnology.   I. Title.
S494.5.B563P74   2003
363.19'2—dc21   2003042823

ISBN 0-7432-2611-9
0-7432-6763-X (Pbk)

To Eleanor,
without whom this, and so
much else of me, would not
have been possible

CONTENTS

Food, Inc.

INTRODUCTION

This book is for those who still have an open mind about genetically modified foods—despite the constant flow of alarms from consumer watchdogs and constant assurances from the agricultural establishment that everything down on the farm is lovely. Those who have already decided what to think should stop right here. This is not a book that will make them feel comfortable. Nor is it intended to persuade them to think differently.

Like other average consumers in this growing debate, I did not set out with strong opinions about genetically modified foods. Nor do my views fit easily now into either camp. I am persuaded that the biotech harvest has considerable perils, if done too fast or without proper regulation, but I can also see that it has considerable promise to relieve pain and hunger for millions of people—if governments, industry, and overzealous sentries don't stand in the way. Too often, it seems to me, the public has been ill served by special interest groups who have sought to promote their products or press their rigid opinions rather than seek the wider interest of humanity. The middle ground, which I shall try to occupy in these pages, was strangely and rather eerily deserted in the summer of 2001 when I first set foot there, and remains to this day somewhat underpopulated.

A decade ago, Americans took their first bite out of a transgenic food. Scientists had found the ripening agent in a tomato that makes the fleshy part go soft, so they flipped the gene upside down and backward, as they put it. The modified tomato then had an extra few days before it started to rot in the normal fashion. The clever idea was to everyone's benefit.

At the time, farmers were picking tomatoes from the vine when they were green and turning them pink artificially with a whiff of ethylene gas. This crude technique allowed the tomato to be picked unripe by machines and travel longer distances, thus making more money for farmers, food carriers, and supermarkets. Consumers were the only losers. The gassed tomato was hard and tasteless. By contrast, the new gene-altered tomato turned red on the vine without going soft, and the farmer had time to pick his crop by machine and get a handsome tomato to market. The new sort of tomato also had a sporting chance of tasting like the garden varieties of yesteryear.

Today plant breeders are still tinkering with tomato genes, but the real push in plant genetics has not been for the benefit of the average consumer's taste buds or nutrition. Instead, biotech companies have concentrated on altering genes in staple crops like corn, potatoes, and soybeans to give them new defenses against pests and allow them to survive being doused by stronger herbicides. These changes have benefited the seed company, the chemical company (often now the same outfit), the farmer, and the food processor.

The biotech industry proudly points to the rapid rise in acreage planted to transgenic crops. A few million acres worldwide were planted with GM (genetically modified) seeds in 1996; by 2002 the acreage had expanded to more than 120 million. But this is still a tiny portion—only 1.3 percent—of the total global cropland, and 99 percent of the total GM acreage was confined to only four countries. The United States grew 68 percent, Argentina 22 percent, Canada 6 percent, and China 3 percent.

Consumers ate corn and tofu without even knowing that they were gene-altered, since the products looked and tasted the same as the older versions and carried no special labels. What had changed in GM foods was inside the cells of the plant itself, but Americans, worrying more about fat content than any adjustment in cell structure, kept on eating the new products. Those who wondered how the agricultural folks were tampering with their foods ultimately put their faith in the Food and Drug Administration to ban any product that was harmful.

Elsewhere the new seeds were greeted with far more skepticism and even barred in some countries after environmentalists raised one alarm after the other. By 1998, the early success of the new technology had begun to turn sour.

The new crops, environmental groups insisted, were indeed different. They were "Frankenfoods" that could cause allergies in humans and mutations in pests. They could produce agricultural monstrosities, such as invasive "superweeds." They could change the ecology of the planet in unpredictable and irreversible ways. They could destroy biodiversity. They could even cause the extinction of important wild plants essential for breeding staple crops. GM crops could destroy America's favorite insect, the monarch butterfly, some warned, adding that these plants could also bring about the elimination of treasured songbirds from European hedgerows. In Europe and Japan consumers became so agitated about the new GM crops that their governments refused to approve the planting of the new crops pending further scientific studies.

The list goes on. Anti-GM forces discovered in taco shells genetically modified corn approved only for animal feed. They alerted the world to windborne GM pollen that threatened organic farms. Some went so far as to suggest that an alien gene used in most of the transgenic crops might cause cancer. The ultimate vote of no confidence in the new technology came toward the end of 2002, when African nations facing starvation turned away U.S. food aid because it contained genetically modified corn that, in their opinion, was poisonous.

To be an ordinary consumer caught in the middle of this turbulent battle was to be deeply confused. Few could unravel the conflicting evidence, as biotech companies desperately tried to sell their new products and ideological environmentalists worked with equal determination to stop them. On top of it all, religious and ethical groups warned of the dangers of competing with God in the garden.

Environmental groups warn that there are no quick fixes for food shortages in the developing world from this new technology and that indeed, biotech farming could make matters worse if used inappropriately. The global North's industrial methods of farming cannot simply be transplanted to Asia, Africa, or Latin America and be expected to work efficiently. But transgenic plants that can grow in poor soils or survive in arid or tropical climates could have considerable direct benefits for the hundreds of millions of people in undeveloped countries who go to bed hungry every night. So far, however, these "miracle" foods have fallen short of the hype.

One example is golden rice, the most famous test-tube plant, which promised to ward off blindness in undernourished children. Although not the instant sure prevention its promoters originally trumpeted, this prototype may eventually lead to plants that can save lives in places that experience dire food shortages—again providing that no obstacles are placed in the way by governments, industry, or special interest groups.

In Africa a parasitic weed of the genus *Striga*, or witchweed, inserts a sort of underground hypodermic into the roots of corn and sorghum, sucking off water and nutrients. On the tiny farms of sub-Saharan Africa, 100 million people lose some or all of their crops to *Striga*. A genetically engineered defense against this scourge is available, but the company has yet to develop it. So far Africans have neither the infrastructure nor the funds to develop it for themselves.

Scientists tinkering with banana genes have come up with a banana tree with big floppy leaves that can resist a devastating airborne fungus called black Sigatoka. But a potential cure sat on a laboratory bench in Belgium for almost a decade.

In the overfed North, the frustrations over biotechnology are different. For a consumer, there is perhaps nothing more offensive than to be kept in the dark about something so personal as the food we choose carefully at the grocery store. Telling stories from the biotech battleground, I have tried to throw as much light as possible onto the safety aspect of these new foods and the inadequacy of the information made available to even the most alert consumer.

Except for committed opponents, experts agree that there is nothing inherently unsafe about genetically modified foods. However, there are possible hazards. Most scientists admit that transferring genes between species is an unpredictable operation that could cause new allergies for future consumers unless proper precautions are taken. Ecologists have argued persuasively about the dangers of spreading laboratory-altered genes into the environment. The pollen of modified crops can contaminate wild relatives on which crop breeders depend for genes to help fight new strains of plagues and pests.

Scientists, politicians, and companies eager to be in the vanguard of the biotech era conceived a system of government regulation designed for a speedy return on the companies' research investments rather than the best protection for consumers. When entomologists discovered that the pollen of America's prized pest-resisting genetically modified corn was fatal to the larvae of the monarch butterfly, the discovery was belittled by the biotech industry. Industry propagandists suggested unwisely that more monarchs were killed each year by collisions with car windscreens than could possibly by affected by the corn pollen. The fact is, the biotech industry had no evidence for this assertion; research on the issue had not been done. In the end the potential of any new technology to harm humans and insects can always be assessed through further scientific study.

Perhaps the most discomforting aspect of plant biotech—and the reason this book is entitled *Food, Inc.*—is the new level of control over food production that the technology has put into the hands of a few international conglomerates. The patent system allots these companies ownership of living organisms essential to food production. And the patents are not just on the product, such as a pest-resistant corn plant, but on each step in the making of that corn. This property right system has resulted in two disturbing trends: it encroaches on the rights of poor farmers in undeveloped countries, and it curbs independent research. Companies in the gene-poor industrial North, mostly from America, have been acquiring traditional plants and herbs from the gene-rich South and then claiming ownership. Scientists working on transgenic crops for undeveloped countries are beholden to the whim of companies with patents on basic laboratory techniques.

Despite lack of consumer confidence, mostly in Europe and Japan, farmers worldwide continued to sow more GM seeds in 2003. The global area planted grew by 15 percent, with an increase of 28 percent in developing countries. The United States accounted for 63 percent of the total, followed by Argentina with 21 percent, Canada with 6 percent, Brazil and China with 4 percent each, and South Africa with 1 percent. Biotech companies continued to explore the horizons of the new technology. In 2003, a new phenomenon known as "biopharming" emerged in the United States—growing pharmaceuticals in corn, rice, and sugarcane because it was cheaper than producing them in factories. Maybe biopharming will someday help poor farmers make a better living and possibly it will even reduce the cost of drugs, but keeping pharma corn separate from food crops posed a new and alarming problem. After a field test, pharma corn that was supposed to have been totally removed was found growing among canola destined for human consumption. Drugs in your French fries, anyone?

Biotech plant researchers kept their trade alive with experiments on a variety of new products from corn with increased vitamin E, a type of cress that can produce the healthy fats normally found in fish, a virus-resistant melon, and a bug-resistant pineapple. They worked on "sentinel" plants that by changing color could alert farmers to pest infestations, plagues, and poor soil conditions—and could also be adapted to detect the presence of chemical and biological warfare agents. In animal and fish laboratories, researchers bred a pig that produced heart-healthy fatty acids and a fish that glowed when it swam into polluted water. In most cases, the companies shied away from developing such exotic inventions because of the regulatory hurdles involved.

But the most significant sign of stagnation in the technology came in the early summer of 2004 when Monsanto, the leading biotech company, shelved its plans to introduce the world's first genetically modified wheat. After spending seven years and hundreds of millions of dollars, Monsanto bowed to consumer opposition to GM products in Europe and Japan by not marketing an insecticide-resistant wheat they claimed had increased yields by up to 15 percent. Other biotech companies abandoned efforts to penetrate the European market with GM corn and sugar beet. American and Canadian farmers had told Monsanto they were worried that if they grew the new wheat, European and Japanese consumers would reject not only the new GM varieties but their entire wheat production. (In the last decade the EU and Japan have bought 45 percent of U.S. exported wheat.) Green groups claimed a big victory while at the same time warning that the war was not over; GM products would still be slipping into the food chain in animal feed over which the consumer had less control.

In the launch of a new technology, powerful forces are always at odds, but this adaptation of the human food supply brought an unusually fierce public reaction. There was ample opportunity to be misinformed and misled. In addition, anyone writing about genetically modified foods faced a basic problem of language. The phrase *genetically modified* is leaden and generally off-putting. So is the acronym *GM*. The word *transgenic,* which seems to be gaining popularity, neatly captures the idea of genes crossing between species, but all of these terms are liable to cause the brain to fog up and move elsewhere. I haven't come up with a solution to this problem, I'm sorry to say. In the book I've attempted to demystify the language, explain the science, and make sense of the scaremongering and the bland assurances. The genetic revolution in agriculture is too important to be left to propaganda, either from corporations or environmental ideologues.

My dispatch starts where the first seeds of contention were sown a century and a half ago, in Gregor Mendel's monastery garden.

New York, July 2004

Mendel's Little Secret

One of the most cherished dreams of plant breeders has been to find a way to transform corn and other cereal grains into super-plants able to reproduce themselves. . . . The term for this type of vegetative miracle is "apomixis."
—U.S. Department of Agriculture press release, 1998

Thinking about how our food is changing at the hands of the genetic engineers leads inevitably to the image of Gregor Mendel, the Moravian monk, breeding peas in his monastery garden a century and a half ago. Dressed always in a black robe, a pair of tweezers in one hand and a camel-hair paintbrush in the other, Mendel bent over rows of peas, cheerfully castrating the flowers by snipping off the pollen-bearing anthers and dusting on a different pollen from another row.[1] He bred round peas with wrinkled peas, peas from yellow pods with peas from green pods, tall plants with dwarf plants, carefully separating each into breeding lines and then crossing and backcrossing them to watch how the traits appeared in future generations.

In time the jolly amateur gardener scooped his fellow nineteenth-century botanists, including Darwin, with his insights into the basic laws of heredity. Mendel was the first to understand that characteristics such as height, color, and shape depend on the presence of determining *factors* (they were not called genes until much later) and that these factors could be either dominant or recessive. For his work Mendel was posthumously acknowledged to be the father of modern genetics.

This popular image, however, misses another, less well known Mendel who becomes important today in the era of genetic engineering. The other Mendel was not so cheerful, a solitary monk still toiling in the monastery garden, but this time struggling without success to comprehend the strange reproductive processes of a common orange-colored wildflower called hawkweed.

In the hawkweed case, Mendel had accepted a challenge from a German professor of botany to crossbreed varieties of hawkweed and figure out what happened to the plant through successive generations. When he had done this experiment with peas, the offspring had shown different characteristics, allowing him to deduce his law of random assortment of the plant factors. The progeny of hawkweed were strangely different. They were all the same in the first generation and continued to be the same in successive generations, bewilderingly exact replicas of the mother plant. Mendel could not figure out what was happening and died, as far as is known, without making any progress in unraveling hawkweed's puzzling reproductive behavior. After his death, all his personal and scientific papers were burned, possibly by a rival monk, in a huge bonfire in the monastery courtyard where his greenhouse had once stood.[2]

We now have an explanation for hawkweed, even though scientists still don't know how it works. Mendel had witnessed a plant that produces seeds without sex, the biological phenomenon of asexuality, known in plants as *apomixis*. Hawkweeds do it that way; so do dandelions. Mendel's basic laws applied to peas and most other living things, but they did not account for the odd behavior of hawkweed.

The word *apomixis* is from the Greek *apo,* meaning "away from," and *mixis,* which means "mingling," a quaint conjunction that aptly describes the somewhat haphazard way plants have sex. Typically, a plant releases a shower of pollen grains that are carried on the wind, or by an insect, to the female organ in the quest to fertilize the eggs. In apomictic plants the pollen is infertile, and the egg itself does all the work. The seed from this activity produces a clone, an exact copy of the mother plant. Instead of having a gene pool constantly changing through the mingling of genes during sexual reproduction, the combination of genes in apomictic plants is frozen, in theory, forever.

Asexual reproduction turns out to be the method of choice for a small but diverse group of plants and animals, from roses and orchids to freshwater flatworms. It occurs in 10 percent of the four hundred families of flowering plants but only 1 percent of the forty thousand species that make up those families. The apomicts, as they are called, include several other wildflowers besides hawkweed and dandelions but only a handful of things we eat, such as mango, blackberries, and citrus.

More than a century after Mendel's death, apomixis remains one of the most vigorously investigated botanical mysteries. Researchers in America, Australia, Europe, and Russia are racing to discover which gene, or combination of genes, governs asexual reproduction. They also want to know whether apomictic plants always produce seeds without having sex. The apomictic dandelion once had normal sex and some primitive species behave like regular sexual plants.[3] Why did they evolve this way?

Oddly, although we now have highly sophisticated techniques for swapping genes from one species to another—powerful laboratory tools and enzymes that snip off the precise pieces of DNA we want to splice—we still have a lot to learn about the sex life of plants.

The best guess so far is that apomixis is a suppression of normal sexual activity. But basic questions remain unanswered about the courtship of plants—how the plant cells send signals to each other during fertilization and whether these signals are different in asexual plants than in plants that reproduce with sex—and what really happens during the formation of the embryo.

Such matters would be of little more than academic interest when it comes to thinking about the future of food except for one important fact. None of the world's major crops is apomictic. When a plant breeder produces a prize variety of, say, corn—handsome, high-yielding, and resistant to pests and plagues—and that corn plant has natural sex with its neighbor, the next generation is always slightly different, just as we are each a little different from our parents. The plant breeder yearns for some method of retaining the most desirable combination of genes in his prize variety year after year.

Apomixis could be the answer—which is why its secrets are known as the Holy Grail of agriculture and why there is a furious international scientific race to solve the mystery. The winner of this scientific trophy could revolutionize agriculture—and harvest massive profits. Apomixis could be of tremendous benefit to seed companies; it could also help the world's farmers, especially those in undeveloped countries.

Since 1935, when the seed companies started selling hybrid corn that lasted only one season, farmers who plant hybrids have been forced to buy new seed each year or fall behind competitors in their production of grain. If those seeds contained the apomixis genes, a farmer would have no need to buy new seed each year because his plants would do as well in the next and successor generations. He would save seed from his harvest, as farmers once did. Apomixis could offer relief for poor farmers in Asia and Africa who cannot afford to buy seed and who still breed their own varieties. They could fix traits in a prized traditional variety. The seed companies would also benefit. Breeding new varieties is a costly and time-consuming business that could be superseded by apomictic plants that fixed their genomes forever.

There is a catch, of course. This promise comes only if apomixis is unraveled by someone willing to share the discovery. If the secret of asexual plants is patented by a corporation that insists solely on commercial gain, farmers in undeveloped countries and most seed companies would be excluded from such an exclusive agricultural club for twenty years at least, the normal life span of an international patent.

In many ways the race to unravel the mysteries of apomixis poses the central dilemma of biotech agriculture. Until now the focus of protests and of the media has been on the taint of new genetically modified (GM) foods, an issue that arises in rich nations where hunger is rare and such food is a matter more of taste than of necessity. While protesters march against "Frankenfoods" and trample on field tests of GM crops, and while the media raise the alarm about toxic GM potatoes and the possible extinction of the monarch butterfly from eating GM corn pollen, both give short shrift to the larger question: how can the promise of this technology and its life-giving products reach those most in need?

The core issue is the increasing dominance of industrial capital over farming, especially in undeveloped countries. If the keys to the creation of the new miracle plants—plants that defy pests, or grow well despite droughts or floods, or produce wonder fruits that serve as medicines as well as food—are locked up in the safe of agribusiness, it's hard to see how poor nations will reap the benefits. If we in the developed world can use a transgenic caffeine-loaded soybean to produce coffee in Minnesota, the coffee workers of Kenya are likely to lose their centuries-old livelihood. If the new technology can help feed the extra three billion people expected on the planet between now and the middle of the century, public funds will have to be set aside to ensure that the technology is available in poor countries. If a new transgenic rice plant can help to cure blindness in those who live on little more than a bowl of rice a day, some new partnership between rich and poor has to be forged so that the intellectual property rights to such a marvelous invention will be shared.

If these inventions are owned by a few international conglomerates, how will these promises be fulfilled? Those who till the world's vast farmlands are in danger of becoming mere contract employees in bailment to a global food processor who supplies the seed with the understanding that the harvest and next year's seed belong to the processor, not the farmer.[4] And we risk having fewer choices even than today in the range of foods we can buy at the local grocery store.

As agricultural science moves relentlessly forward, some enlightened new private and public partnerships are emerging so that these technological advances have a chance of being shared. In theory, the new arrangements take into account the needs of different farming systems in different countries, but will they allow farmers to grow their favorite and traditional crops rather than homogeneous foods for the conveyor belt of industrial agriculture? The fear of those opposed to the new technology is of a "plague of sameness," a vast monoculture organized and guarded by some big brother corporation.[5]

These are not new issues. They have been around for a hundred years, since the application of Mendel's laws of heredity slowly turned crop breeding from a rural art into a science.[6] However, the issues came into sharper focus on the eve of World War II when the yields of the new hybrid corn varieties were outpacing anything that had gone before, and when the means of agricultural production, the seed, began changing hands, from a public resource like air to private ownership. Swarms of John Deere tractors started plowing up the American Midwest and any foreign field where farmers or nations were rich enough to purchase the machines. Tons of artificial fertilizer were spread on those lands, clouds of new powerful insecticides and pesticides were sprayed on the bounty, and the harvest was brought home with mechanical pickers to stock the industrial world's grocery marts.

In 1962 came the counterrevolution. Rachel Carson protested the devastating effects of these chemicals in her book *Silent Spring,* which led to a new public awareness that forced chemical manufacturers to restructure the formulae of their toxic wares. But the high yields were too important, and industrial agriculture marched on, using different chemicals that helped produce so much food that farmers entered a vicious spiral of overproduction.

In developed countries during the last half of the twentieth century, the average crop yields of wheat, corn, and rice doubled or tripled, the number of tractors in the world rose from seven million to twenty-eight million, and the average annual yield of a milking cow in France increased from fewer than two thousand liters to more than five thousand. The production increases drove down prices paid to farmers, while farmers' costs rose. The loss of the family farm became the sad anthem of rural America as the nation and the rest of the developed world shifted to industrial agriculture.

This farming revolution passed by most of the world's farmers, who, being poor, continued to use manual tools and raise crop plants and animals that benefited little from the intense breeding of improved varieties.[7] The gap between the most productive and the least productive farming systems increased twentyfold.

By the 1980s the biotech agricultural revolution was brewing. The application of genetic engineering to crop plants, by allowing a desirable gene from one species to be inserted into another species, offered agribusiness a new method of control. The chemical company that sold powerful, all-embracing new weed killers now also sold seeds that grew into plants especially designed to resist those herbicides. To compete, farmers had to buy both seeds and weed killer. Once again, only those who could afford the new package survived. The improvements never reached the poorest farmers in Africa. The seed companies were not interested in producing pest-resistant cassava for farmers who would not be able to pay for it.

With the appearance of the first genetically engineered whole food—a tomato that didn't rot on its way to market—a food war broke out between agribusiness and a diverse group of activists in the developed world. Scientists, doctors, environmentalists, ecologists, farmers, agronomists, sociologists, lawyers, economists, creationists, mystics, latter-day Pre-Raphaelites, and antiglobalists who wanted to bring a halt to this new technology took to the streets to stop agribusiness from tampering with their food.

But the antibiotech forces were not urging scientists and companies to tailor their genetic inventions in ways that could help the millions of hungry people in the world. There were no banners urging "Miracle Seeds for the Poor" or "Gene-Altered Cassava for Dry African Fields." Some protesters demanded nothing less than a halt to the "unnatural," even ungodly, practice of swapping genes between species. Their argument was not that genetic engineering might be put to better use, but that it was of no use. They focused on the scientific possibility that the new foods could be unsafe, that they were an unnecessary experiment perpetrated by scientists without a social conscience and wicked corporations intent only on profit. They worried that transferring genes between species might cause allergies, or worse; alien genes might "escape" into the wild and create "superweeds" and "superpests" that could disrupt the world's ecosystems.

In Europe the British government was reeling from food scandals, the contamination of pork and poultry with dioxins, and the "mad cow" epidemic. The battles reached such a pitch that the Europeans banned imports of the new transgenic grains except for animal feed and demanded that all products containing the new foods be labeled. The Japanese banned imports of the new modified corn. As a result U.S. farmers lost important markets and became uncertain which seed to plant next season. The food industry panicked and, fearing they would be unable to sell their famous brands abroad, demanded that suppliers provide grains free of genetic "contamination." Looking at the agricultural casualty list in 1999, an analyst in the New York office of Deutsche Bank declared, "GMOs [genetically modified organisms] are dead."

The attitude of the agribusiness companies did not help. They were as arrogant about their new "miracle" foods as the nuclear power industry had been about the "peaceful" atom in the 1960s. The biotech scientists in the big universities made the same mistake. They boasted, "We've invented fire. The sky's the limit," an uncomfortable reminder of the forecasts of their atomic colleagues who promised that electricity would soon be clean and risk free.

Governments, scientists, and companies thought that they could rally public support behind the new technology without informing citizens of the true nature of biotechnology. But agriculture is different from other sectors of the economy, such as drugs and cosmetics. Rural life has always held a special place in any nation's cultural heritage, in its cuisine and in its art. Think of the farm scenes of Bruegel and Constable, for instance. Although much of this feeling is a misplaced nostalgia for supposedly idyllic life that is, in fact, quite beastly, farming is not merely a job, it is also a mission. Bringing food to people's tables not only provides for others but also encourages the roots of self-sufficiency and community. The land is where any nation cares for its economic, social, and environmental health, the place where ecosystems, biodiversity, and water quality are nurtured.

In the war over genetic agriculture, the public soon demanded more debate. Prodded by green groups, biotech companies found themselves explaining and defending the right to experiment with complex aspects of genetic engineering that they had imagined were safely secreted in agricultural laboratories. Their view was that the public could not be bothered with and did not really need to know about antibiotic marker genes, the cauliflower mosaic virus, the gene flow in Mexican corn fields, and jumping genes that might under certain circumstances create new allergens and toxins. All these matters were to be avoided as far as possible as the stuff of public discourse. It was a colossal miscalculation, and when the public caught on, the result was widespread confusion and alarm.

Whether it takes twenty months or twenty years before scientists break the genetic code for apomixis, that day will surely come. And then plant breeding will finally enter its next phase. Scientists will have moved beyond the simple transfer of one gene to another to make crop plants short or tall, or to increase a plant's own defenses against insects and pests, or to bestow resistance to cold or heat.

In the year 2000, scientists took a step into that new era. Two German researchers used three alien genes, two from a daffodil and one from a bacterium, to create in rice a substance known as beta-carotene. Under the right conditions, beta-carotene can be converted in the human body to vitamin A, which is missing in the diet of millions of poor people, causing blindness and defective immune systems. The new rice turned yellow, like a daffodil, and was instantly dubbed "golden rice."

This book enters the debate over genetic agriculture at the point when those two German scientists created golden rice, a "miracle" crop by any standard. Golden rice created a possible change in the food supply that Mendel could not have fathomed from his monastery garden, any more than he could comprehend the strange sexuality of hawkweeds and dandelions.

What seemed like a noble humanitarian effort, however, quickly turned into the loudest battle of the biotech wars.

2

SEEDS OF GOLD

> It turned out that whatever public research one was doing, it
> was in the hands of industry and some universities.
>
> —INGO POTRYKUS

One February night in 1999, Professor Peter Beyer was working late in
his laboratory at the Freiburg Botanical Gardens in southern Germany. Before going home, he decided to take one more look at the
grains of rice sent for analysis by his colleague in Zurich, Ingo
Potrykus. He carefully removed the grains from the polishing machine
and held them against the light. Instead of the usual pearly white, they
had turned a beautiful translucent yellow. Trying to contain his excite-
ment, he checked the data. The chemical analysis confirmed that the
color in the grains was no trick of the eye. For the first time in agricultural history, a grain of rice, the world's most important food crop,
contained the pigment beta-carotene, the same substance that turns
corn yellow and carrots orange. Beyer went to his computer, typed in a
message about the success of the experiment, and e-mailed it to
Potrykus.

One of the most exotic and promising ideas to come from the
botanical laboratories of the genetic revolution had become a reality.
Beta-carotene, an essential nutrient for the human body, produces
vitamin A. The lack of this vitamin causes the death of an estimated
1 million of Asia's poorest children each year from weakened immune
systems. Another 350,000 go blind.

Although beta-carotene is present in the leaves of the rice plant and the husk of the grain, which is removed during milling, the pigment had never found its way into the rice grain. Potrykus and Beyer had spent six years coaxing reluctant rice plants to complete the complex chemical reaction that had not occurred naturally in more than ten thousand years of human cultivation.

The tiny yellow grains in Beyer's laboratory would quickly become the most visible invention of the new transgenic food industry. Agribusiness companies that had invested billions of dollars in the future of such technology hailed golden rice as a tremendous victory in the war over genetically modified foods, a fine example of how this new generation of biotech foods would save the world from starvation and malnutrition. Opponents quickly took up their own war cry, denouncing golden rice as a corporate hoax. For them, golden rice would become a rallying slogan in their struggle to put an end to the use of genetic engineering in agriculture.

In all the hubbub, no one denied that Potrykus and Beyer had indeed pulled off a dramatic scientific coup, confounding the predictions of peers who had forecast failure. They had isolated three genes—two from a daffodil and one from a bacterium—and inserted them into a rice plant's elaborate genome, which contains about fifty thousand genes. In the process, the two scientists had "instructed" the plant to complete a chemical chain reaction ending with the production of beta-carotene in the rice grain.

But the politics of the experiment were even more significant than the science. Here, at last, was an invention that the biotech industry could be as proud of as the critics could find alarming, or so it seemed. Golden rice put a humanitarian face on a technology that had been producing strange new foods without regard to the nutritional needs of the customer. The first generation of transgenic plants had benefited three groups—farmers, seed merchants, and food processors. For them, a tomato that didn't rot on the way to market was a botanical miracle, as was a soybean that could live through sprayings of tough new herbicides. Corn plants with built-in resistance to pests promised to save on pesticides, just as a canola plant that produced more oil boosted the bottom line. But the consumer never noticed the difference. Now golden rice moved the new technology beyond the world of agricultural production, raising hopes of both the industry and the hungry that it could benefit millions of people—and not well-off, overfed Western consumers but poor, undernourished people in Asia and Africa.

The biotech industry seized the moment, launching a TV advertising blitz that trumpeted the possibilities to end world hunger and disease. Smiling Asian children were pictured being nurtured by caring doctors against a backdrop of rice paddies under the headline, "Save a Million Children from Going Blind." *Time* magazine ran a front cover declaring, "This Rice Could Save a Million Kids a Year," over a picture of Potrykus in his greenhouse peering out from behind his wondrous golden grains.[1] Even President Bill Clinton joined in the celebrations. "If we could have more of this golden rice . . . it could save four thousand lives a day, people that are malnourished and dying."[2] The journal *Science,* which was the first to announce the successful experiment, distributed copies to seventeen hundred journalists around the world with a commentary expressing hope "that this application of plant genetic engineering to ameliorate human misery without regard to short-term profit will restore this technology to political acceptability."[3]

Golden rice, the companies promised, was just the start of many more transgenic crop plants for the underprivileged—bananas with vaccines against tropical complaints, corn plants that produce pharmaceuticals more cheaply than factories, plants that would produce not only vitamin A but a veritable pharmacy of the basic nutrients needed for a healthy life.

For the inventors, golden rice offered a fast track to international fame (and possibly more research funds). Dr. Potrykus counted thirty TV broadcasts and three hundred newspaper articles devoted to golden rice in the first year. The media turned the two scientists into instant heroes, not without reason. They had labored in government laboratories, always short of funds, pursuing the kind of project the companies had rejected because there were no profits to be had from the poor. Potrykus became a roving ambassador for golden rice, urging companies and governments not to waste "even a single month" in pushing his invention into production; the alternative was too horrific to contemplate: "Fifty thousand children will go blind [this year]," he warned.

To the opponents of genetic engineering, the golden grains in Beyer's laboratory were "grains of delusion" and "fool's gold."[4] The two scientists were corporate dupes, trapped in the folly of "industrial agriculture." Certainly, if golden rice were ever to be an effective weapon against malnutrition, it would have to be grown on millions of acres. Such monocultures, the critics argued, encouraged crop failure, destroyed traditional varieties, favored the rich at the expense of poor farmers, and put the production of the world's food supply into the hands of a few. The spectacular failures of monocultures were well known. More than a million people starved to death in Ireland in 1845 because of the blight that rotted an entire season's monoculture crop of potatoes. More than a century later, another blight hit the cornfields of America when certain widely used hybrids in 1970 produced a scant half of the projected yields. Monoculture encouraged farmers to abandon their traditional varieties and plant "miracle" crops; the practice threatened the survival of seeds that had been carefully cultivated over centuries. Without these *landraces,* or heritage seeds, it would also be impossible to pump new genetic life into crops to fight off plagues and pests.[5]

Potrykus, the leader of the team, might even be an industry mole, critics suggested, noting that he had worked as a molecular biologist for Ciba-Geigy, the Swiss seed company, and had been granted several patents of agricultural interest, at least one of them for "transforming" plants with alien genes.

Golden rice was a Trojan horse, the protesters declared, a secret weapon designed only to pump life back into a new technology struggling to survive. The timing of the appearance of golden rice was "so clear," said Greenpeace, seeing corporate conspiracies at work. As one of the group's leaders put it, "People are talking about the potential benefits of the second generation of genetically modified crops when almost no questions raised by the first have been answered. You don't have to be paranoid to think the tactics are deliberate."[6]

To the green activists, golden rice was another technical fix being promoted as the solution to the problem of malnutrition when the world was already awash in food and vitamin pills and pharmaceuticals. The problem, they argued correctly, is that people were sick and hungry not because of global shortages but because of wars and dictators and simply because poor people could not afford to buy what was already available. For these detractors, the Nobel Prize–winning economist Amartya Sen's famous study of poverty and famine explained how access to food depends on a complex mix of economic, social, and political factors, how a poor person may easily starve in the face of plenty. Instead of trying to develop another new high-tech rice, they suggested a lower-tech solution—distribution of vitamin A pills to those in need.

Potrykus and Beyer, they asserted, were the tools of a life-science oligopoly taking over the world's food supply. If the new rice were ever produced, its only function would be to boost the profits of the agribusiness conglomerates who sold the seeds and owned the patents.

In each of these arguments there were grains of truth, but there were also distortions. Like the biotech companies, the green activists angled their critique to their own advantage. Golden rice had arrived in the middle of a propaganda war that the companies had lost in Europe and were in trouble with in America. In Europe, Greenpeace and Friends of the Earth were swelling their ranks with "Frankenfood" slogans. Prince Charles was rallying the God-fearing against what he and his followers saw as the ungodly act of transferring genes across species.

In America, groups such as the Union of Concerned Scientists, Environmental Defense, and the antibiotech activist Jeremy Rifkin's Foundation on Economic Trends were putting up mostly responsible challenges to an industry that was behaving irresponsibly in not explaining the details of its new technology to a wary and suspicious public. For a while the future of genetically engineered foods looked uncertain in the developed nations, if for no other reason than that no one could quite see the benefit of the product to the consumer.

To anyone living in poverty in, say, the Indian subcontinent, Indonesia, Vietnam, Thailand, or the Philippines, where there are the most severe shortages of vitamin A, the benefit, if true, was seen from a much different angle. However, poor people in these regions should be forgiven for seeing the moment of celebration in Freiburg's botanical gardens as a public relations stunt. Since the first gene-altered whole food had appeared on the market in 1994, not one product had come their way. Bioengineering of crops had passed them by. None of the big agribusiness companies—the Swiss Syngenta, the American Monsanto or DuPont, and the German AgrEvo—had produced improved varieties of rice, cassava, or yams, the staple crops of the third world (although Monsanto had developed a high-beta-carotene mustard plant that it decided to give free of charge to poor and subsistence farmers).[7]

The only way undeveloped countries came into contact with the products of the first generation of the biotech agricultural revolution was through food aid—millions of tons of genetically modified corn from the U.S. corn belt that American farmers were having trouble selling to their usual markets in Europe and Japan. African countries such as Zimbabwe and Zambia, alarmed by the rhetoric from Greenpeace and Friends of the Earth about the risks attached to the new foods, were wary of taking the corn. Either, they thought, the corn was unsafe to eat, or more importantly, their farmers might plant some of the seeds and the alien genes might somehow escape and "contaminate" traditional native varieties.

The anti–golden rice forces told Asians and Africans there was nothing worthwhile in this golden rice, not even the beta-carotene in its grains. Greenpeace claimed that no one could eat enough grains to supply the daily intake of vitamin A. The group estimated that as much as twenty pounds of cooked golden rice a day would be needed to meet the daily requirement of vitamin A. In many undeveloped countries, a pound of rice per person each day is a luxury.[8] Even if scientists increased levels of beta-carotene in the rice, people eating it needed enough fat in their bodies to complete the chemical reaction from food to vitamin. Unlike overfed Westerners, most poor people have little chance to add fat to their meager diets. At best those first grains could contribute only between 15 and 20 percent of a human being's daily vitamin A requirement, golden rice promoters conceded.[9]

Still, nothing was straightforward about turning beta-carotene into vitamin A, the naysayers cautioned. The beta-carotene could be degraded, or even destroyed, by exposure to light and during processing, heating, and storage. For beta-carotene to survive cooking, it's best to microwave or steam the foods, not boil or sauté.[10] Such culinary niceties were unlikely to be available in undeveloped countries, they said.

Few critics rivaled Vandana Shiva, a fifty-year-old Indian scientist, feminist, and international campaigner against the new technology, who beat the drums against the evils of corporate agriculture from her own small research institute in New Delhi. Books and pamphlets on display there included *The Violence of the Green Revolution, Biopiracy, The Future of Our Seeds,* and *Monsanto: Peddling "Life Sciences" or "Death Sciences."* Among the books sat jars of traditional Indian rice varieties.[11] Appearing at international conferences on biotechnology in her flowing orange saris, Shiva had become a cult figure in the war, beloved by the antibiotech forces for her ability to ignite an audience's passions with her Gandhian-style rhetoric. Loathed by the biotech company executives—and some biotech scientists—who watched the same audiences with alarm, Shiva railed against Western industrial agriculture. The muscular agriculture establishment was systematically destroying the alternative crops of fruits and vegetables that were a much better and more available source of vitamin A than any rice produced in a Swiss laboratory.

Shiva became especially incensed as she told people about the demise of *bathua*, a popular leafy vegetable rich in vitamin A and grown in North India. She charged that bathua had been pushed to extinction by herbicides used on new Western varieties of wheat. "The 'selling' of vitamin A as a miracle cure for blindness is based on [the corporations'] blindness to the alternatives," she said. One alternative was to persuade people to produce more greens in their home gardens. "You don't live by corn alone. You need the squash and you need the fruits," she said.[12] Indeed, home-based gardening had been a popular strategy in Asia for the control of vitamin A deficiency. Studies had demonstrated that radio spots and other educational programs, supported by the media, did result in the increased cultivation of the right vegetables.

Even if golden rice were to overcome the technical, safety, and environmental issues, who would eat yellow rice? critics wondered. Even the poorest people who live on rice prefer the white grains as a status symbol, often with religious significance. People could be as fussy about white rice as Europeans used to be about white bread, or American Southerners before the Civil War were about eating white corn. They gave their slaves yellow corn, not realizing that yellow corn was more nutritious because it contained beta-carotene.

In fact, rice comes in a rich variety of colors. Beer and spirit makers in Myanmar prefer the black rice that ranges in hue from midnight to deep purple. Red rice is quite popular in Madagascar, in Iran and northern Ghana, and was once widely grown in Taiwan.[13]

The final argument against golden rice focused on the rice plant itself, a variety known as *Taipei 309*, a short-grained *japonica* type suited to northern temperate zones but not found in the tropics or eaten by the poorest people. The scientists had chosen *Taipei 309* because it was easier to perform the experiment on *japonica* than on *indica* varieties. No one knew how golden rice's new combination of genes would behave in *indica* rice varieties common to tropical and subtropical zones where great numbers of people suffer from hunger and malnutrition.

By the time the critics had done their work, agribusiness claims about golden rice looked absurdly overblown. Even the original sponsor of golden rice, the Rockefeller Foundation, was forced to admit that the prospect of immediately saving the sight of half a million children had been exaggerated; for the moment, they had only a research project several years away from producing a viable crop.

Potrykus was taken aback by the storm his golden rice had created. After the previous three years of the biotech war, he had expected opposition, some "poisoning of the springs," as he put it in the German phrase Brunnenvergiftung, but he found the wave of abuse and criticism overwhelming. In an article in the journal *Plant Physiology*, he bemoaned "the propaganda war against our work with arguments that we are only pretending to work for mankind, or are only satisfying our own egos, or are merely working for the profits of industry." [14] Did his critics not realize, he asked, that his research team was also working on boosting the iron content of rice, which contains less iron than other cereals? Didn't they know that 1.4 billion women suffer from iron deficiency? [15] (This experiment, which was completed a year later, succeeded in doubling the iron content of the same *japonica* rice variety.)

Like many of his colleagues working in genetic laboratories throughout the developed world, Potrykus had desperately wanted to be part of the "biotech decade." For him, however, time was pressing. He had grown up in Germany during World War II. His father, a medical doctor in the German army, had been killed when an Allied bomb hit his troop train. The family had struggled after the war, but Potrykus was a good student and quickly made his way in the new biological sciences. He had been a full professor at the Institute of Technology in Zurich since 1985, but his employer, the Swiss government, insisted on retirement of its employees at the age of sixty-five. In 1999 Potrykus would automatically lose his laboratory and his researchers. As he viewed this finality from the perspective of the early 1990s, he realized that a decade was not a long time in the competitive field of crop science.

In his final productive years, he could have used public funds, as he would later observe sarcastically, in a "nonpolitical experiment to study why the hairs on the leaves of the small weed *Arabidopsis thaliana* are sometimes two and sometimes three-forked." [16] But Potrykus was interested in breeding cereals, trying to make crop plants more nutritious—especially rice, the primary food for more than a third of the world's population. Cereal crops were the big challenge for agricultural scientists; moving alien genes into tobacco and petunias was relatively easy; modifying cereal crops was proving far more difficult. By all accounts Potrykus was a demanding professor, never content with mere encouraging indications of success; he wanted proof. "Potrykus was the evil empire," recalls one American scientist, referring to his tyrannical demands for details of experiments. The list of "Potrykus's postulates"—the steps that had to be overcome before he would accept that a gene had been transferred into a host plant—provoked both fear and respect in the agricultural community. [17]

The biotech companies had no interest in strategies to fight malnutrition in developing countries; profits were in the gene-altered herbicide and pest-resistant crops for the big farms of North America and the prairies of Argentina and Brazil. But the United Nations and several international aid agencies were very interested. Despite big improvements in global food supplies since the '60s, more than two billion people, especially women and children, lacked sufficient vitamins and minerals in their diets, particularly vitamin A and iron. In 1990 heads of state at the World Food Summit for Children set a goal of eliminating vitamin A deficiency by the year 2000. Several aid programs were launched to distribute vitamin A capsules, to add iron to wheat flour, and to educate poor people about their diet. But distribution was uneven and often impossible in remote areas. The UN's Food and Agriculture Organization, the World Health Organization, and leading food research groups suggested that the real solution was to increase the amount of the missing nutrients in the staple crops.

Potrykus searched in vain for private funds for his rice transformations. None of the private companies was even studying rice. At the nearby University of Freiburg, Beyer was studying enzymes (catalysts in chemical reactions) that regulate the metabolic pathway that produces carotenoids, the family including beta-carotene, in daffodils. Chemical reactions in cells produce *metabolites;* inside the cell, metabolites either break down to create energy to keep the cell alive or build up more complex substances, such as beta-carotene. The small steps taken by individual chemicals to create these more complex substances is known as the *metabolic pathway.* Beyer wanted to see if the enzymes involved in producing carotenoids in daffodils could be inserted into the rice plant to create the same pathway in the rice grain. But he also needed funds.

In 1985 the Rockefeller Foundation, a longtime funder of agricultural programs for developing countries, committed half of its funds earmarked for agriculture—more than $100 million—to rice biotechnology. The foundation trained Asian scientists at advanced Western laboratories to produce rice varieties that would benefit poor farmers and consumers. A perennial question was, Why couldn't the gene or genes that turn a kernel of corn yellow with beta-carotene also work in the rice grain?

Early research showed that four enzymes absent in the rice grain were needed before beta-carotene could be produced. At Iowa State University, researchers isolated and cloned a gene from maize responsible for the production of one of the four enzymes. But where to find the other three?

In 1992 Rockefeller held a seminar in New York of their beta-carotene pathway experts and invited Potrykus and Beyer, who offered his daffodil genes. The experts were skeptical. Potrykus himself admitted, "There were hundreds of scientific reasons why the introduction and coordination of these four enzymes would not work [in rice]—and some concern that there might be problematic side effects." One of the enzymes had to combine with a chemical compound in the rice grain that also played a crucial role in the germination of seeds and the development of seedlings. Potrykus and Beyer were worried that borrowing from this valuable store of chemicals might reduce the fertility of the seeds. (In the experiment, it had no such effect.)

But Potrykus was not going to be put off by the naysayers. "With my simple engineering mind, I was naïve enough to believe that it would work," he said.[18] Rockefeller was impressed with his boyish enthusiasm and put up one hundred thousand dollars, a token sum in research terms. In Freiburg, Beyer isolated three daffodil genes he thought might produce the beta-carotene pathway. One of the daffodil genes turned out to be unusually complex and difficult to work with, so he substituted an enzyme from a bacterium. He added regulator genes, or *promoters,* designed to switch the others on at the right moment, and sent the genes to Zurich.

Potrykus was the transformation expert. His laboratory was one of a dozen around the world trying to transfer alien genes into rice plants. Since the mid-1980s, it had been possible, in theory, to transfer any gene across species to any other organism. To succeed in a crop such as rice, the new gene must become an integral part of the new transgenic plant without upsetting the plant's ability to produce edible seeds.

By the early 1990s scientists had tried several methods: adding genes to *naked* plant cells (ones whose tough, thick walls had been removed so that the DNA could penetrate the cell), injecting (firing DNA-coated pellets into the plant), or using a vector such as a bacterium to ferry the gene into the plant.

At first the Zurich team tried to insert the genes one at a time, hoping to combine them later by conventional crossbreeding. But the experiment was a failure. Only the ferry method worked. A common bacterium found in soil *(Agrobacterium tumefaciens)*[19] transfers its genes into plants by invading a wound, usually caused by an insect or wind blast, in the plant's stem or roots. The bacterium then forms a tumor on the plant and its genes mix with the plant's genome. The tumors commonly attack only the big group of flowering plants with broad leaves, including potatoes and tomatoes; when tried with narrow-leafed crop plants, such as maize and rice, the bacterium was reluctant to help.

By the summer of 1998 several researchers had solved the problem. Beyer's daffodil genes were successfully transferred into the rice. Potrykus began growing the seedlings in a bombproof greenhouse in a village outside Zurich. Protests against transgenic food were being staged all over Europe, including the ripping up of test plots, so the glass on the greenhouse was thickened so that it could withstand a grenade attack. In December, Potrykus sent the mature plants to Beyer for analysis.[20]

When the positive results were confirmed, the two scientists planned to send their golden rice seeds to the world's leading public rice research establishment, the International Rice Research Institute, in the Philippines, for adaptation and distribution to farmers. (IRRI, founded in 1959 under an agreement between the Philippine government and the Rockefeller and Ford foundations, had been responsible for the Green Revolution's improved rice varieties.) But Potrykus and Beyer were already well aware that the independence of the project had been compromised.

In the middle of the project, they had run out of money. The Rockefeller funding had dried up after $600,000 over six years. Beyer applied for funds from the European Union, but it was only available for research beneficial to Europe and on the condition that Beyer take on a European industrial partner, who would have options on the product.[21] Rice was not a high-priority crop in Europe, so Beyer extended his research program to include work on a metabolic pathway for making beta-carotene in potatoes. And he took as a partner the British pharmaceutical giant AstraZeneca, which would merge its agribusiness with Novartis to become Syngenta. The EU gave Beyer $240,000 and the Swiss government matched the funding with a grant to Potrykus. The funds paid for one postdoctoral researcher and two Ph.D. students each. "Not much," observed Beyer.[22] The total for the project was now over $1 million.

The only part of golden rice that came without condition attached was the *japonica* rice plant itself, the variety known as *Taipei 309,* which Potrykus had obtained from the IRRI rice establishment in the mid-1980s to carry out his preliminary experiments on transferring alien genes.

Far from putting a failing industry back on its feet, golden rice propelled the war against the technology beyond the media buzzwords of "Frankenfoods" and "superweeds." Critics began to pick at the details of the process of genetic engineering with the spotlight on the content of the *gene cassettes,* the packages of genes including promoter and marker genes that were being inserted into the host plants and which critics argued posed unacceptable risks to human health.

Within a few months of its announcement, golden rice looked a lot less appetizing, but the final blow to its prestige had nothing to do with its color, taste, or nutritional content. The question was, who owned golden rice? This was a new invention in a highly competitive industry riddled with patents protecting billions of dollars invested in laboratory research. Could anyone ever give it away free of charge to the poor, as Potrykus and Beyer had always said they would?

When academic researchers carry out experiments, they never know precisely the extent of the patents covering their work. New patents are continually being issued, older patents expire, and patents may be challenged in court anywhere in the world. Between fifty and a hundred new plant biotech patents and applications for patents are issued each month for each step of the process and the patent portfolio relating to any one experiment is constantly changing.

Potrykus said he had had to ignore the patents while he was doing the experiment, "or I couldn't move at all." He suggested that all scientists should do the same to help spread the benefits of genetic engineering to undeveloped countries. "What company wants the negative publicity of putting me in jail for fighting poverty?" he asked.[23]

When experiments are successful, it is customary to carry out a professional patent search to find out who owns what. In the case of golden rice, the search was complicated by some broad claims skillfully written by patent attorneys to cover as many steps as possible of the transformation process. Some patents overlapped, with two or three covering basically the same technique. For example, Monsanto had been working on boosting levels of beta-carotene in rapeseed, which produces canola oil. The company had a catchall patent covering "a transgenic plant which produces seed having altered carotenoid levels." Although it was a different experiment, using different genes put into rapeseed, not rice, the patent as written seemed to apply.

The liability for the user was often hard to assess. Everyone was stunned to find out that Potrykus and Beyer could have infringed a grand total of seventy patents belonging to thirty-two different corporations, whose permission would have to be granted before any golden grains could be given away.[24] All the big agribusiness companies had a stake in golden rice through intellectual property rights of one kind or another (DuPont, Monsanto, and Zeneca), plus a number of universities (Maryland, Stanford, Columbia, and California). The daffodil genes were covered by patents held by Amoco, DuPont, Zeneca, and ICI. Japan's Kirin Brewery held patents on the bacterium used by Beyer and Japan Tobacco on gene transfer.

Golden rice was a disturbing example of just how close the seed conglomerates had come through international patents to owning every step of the process of taking a gene from one plant and inserting it into another, as well as the transgenic product containing the alien gene.

As the extent of the private ownership of golden rice was revealed through an examination of the patents, attacks on the inventors intensified. By this time Potrykus had become quite good at honing his public outrage. "I was upset," he wrote in a science journal. "It seemed to me unacceptable, even immoral, that an achievement based on research in a public institution and exclusively with public funding and designed for a humanitarian purpose was in the hands of those who had patented enabling technology earlier. . . . It turned out that whatever public research one was doing, it was in the hands of industry (and some universities). At that time I was much tempted to join those who fight patenting."[25] But to what avail? Potrykus already knew that his colleague Beyer had received funds from the EU, and those funds meant that "golden rice" was already tied to the British company AstraZeneca.

Under the terms of the grant from the European Union, the British company had "nonexclusive rights" to golden rice. Potrykus and Beyer could not give away their precious invention to poor farmers in Asia unless AstraZeneca agreed. Neither scientist had personal funds to continue the next development phase of golden rice—transferring the daffodil genes to rice varieties consumed in Asia, a laborious and expensive business.

Unless AstraZeneca waived its rights, which the company was not willing to do, Potrykus and Beyer had to make a deal. To strengthen their hand they decided to patent the part of the invention that was exclusively theirs—the creation of the metabolic pathway. The patent was attained through a small rights company attached to the University of Freiburg. The two scientists then struck a deal with Astra-Zeneca that assigned them rights to give golden rice seeds to farmers in developing countries who earned less than ten thousand dollars a year, and the company the right to market golden rice in developed nations, particularly the United States and Japan. If anyone in those countries was interested in buying golden rice, the company could make a profit. The company agreed to do the necessary research related to environmental and health issues in preparation for government regulatory hurdles in the countries where golden rice would be marketed—probably not before 2006. They set up a Humanitarian Board to oversee the development phase, and whatever they discovered about golden rice in their efforts to pass regulatory inspections in the United States and Japan, they agreed to make freely available to Potrykus and Beyer. Some people saw the golden rice deal as a possible blueprint for the future: a new type of partnership between public science and private industry for the benefit of millions of poor consumers.

As soon as the deal was announced, the other corporations who held patents on materials or methods used in golden rice also waived their rights—rather than be seen trying to block a deal that could "save a million kids a year," as *Time* put it. The day the issue of *Time* magazine hit the newsstands, Monsanto called to donate its patented material. Potrykus was delighted. "I consider the Monsanto offer important because I can now use this case to tell other companies, 'Look, Monsanto is giving me a free license. Won't you do the same?' What could be better?" The compromise they struck gave the beleaguered ag-biotech industry the opportunity they were seeking to become the good guys for a change.

But by this time the unfortunate Potrykus could do no right in the eyes of his critics. The deal with AstraZeneca brought accusations of surrender. Golden rice was "a case study in public science's failure to understand and address patent issues." The two German scientists had been "golden goosed." [26] The critics pointed out, correctly, that only a handful of the seventy corporate patents were relevant to further development of golden rice in Asia, and they wondered why Potrykus and Beyer had not put off filing their own patent until all the companies had given in. Then they could have published instead of filing for a patent. That way golden rice would have remained a public venture. But the two scientists maintained that their hands were tied by the agreement with AstraZeneca. Several critics pointed to the curious position of the Rockefeller Foundation, whose charitable donations had funded more than half of the project with the aim of helping the poor and ended with the rights to the product shared between the inventors and a commercial enterprise. But in fifteen years of biotech funding, Rockefeller had never created any legal mechanism for ensuring that the products of their research went exclusively to the poor. If anyone had been goosed, it was the foundation.

However reasonable or unreasonable the arrangements with Astra-Zeneca looked in the end, the extraordinary intrusiveness of plant patents as they affected new biotech crops had been fully exposed. Immediate demands for their reform opened up another front in the war.

The only part of golden rice that went without further challenge was the rice plant itself, *Taipei 309*. The variety was a product of the Green Revolution—the program of crop improvement in Latin America and Asia launched at the end of World War II, sponsored by the Rockefeller and Ford foundations and guided by the U.S. government. Although the program was stunningly successful in increasing grain yields, thereby averting mass starvation on the Indian subcontinent, the Green Revolution had come under severe criticism for imposing on the global South the industrial agricultural systems of the North and was now a prime target of the antibiotech forces.

To Vandana Shiva, the Green Revolution was the beginning of all the agricultural problems of the undeveloped world. "Instead of millions of farmers breeding and growing thousands of crop varieties to adapt to diverse ecosystems and diverse food systems, the Green Revolution reduced agriculture to a few varieties of a few crops (mainly rice, wheat, and maize) bred in one centralized research center in the Philippines for rice and Mexico for wheat and maize." Instead of solving the problem, golden rice had merely masked the shortcomings of the Green Revolution.[27]

The Plague of Sameness

U.S. agriculture is impressively uniform genetically and impressively vulnerable.

—U.S. NATIONAL ACADEMY OF SCIENCES, 1972

Exploitative agriculture offers great possibilities if carried out in a scientific way, but poses great dangers if carried out with only an immediate profit motive.

—M. S. SWAMINATHAN, 1968

Our ancestors, in their hunter-gatherer phase, helped themselves to thousands of different kinds of fruits, berries, grains, leaves, nuts, and roots, but when they settled down to the grueling business of farming about ten thousand years ago, inevitably they cut the menu. Fewer than two hundred plants were eventually selected for cultivation. Staples came mainly from the grass family, Gramineae, the largest in the plant kingdom. Wheat and barley were grown first in the Fertile Crescent, rice in Asia, the feathery seeds of sorghum in Ethiopia, and maize in Mexico. In South America people cultivated root crops—tiny, knobbly potatoes in Peru and cigar-shaped tubers of cassava in Brazil. As seeds were selected from the most suitable plants, crops slowly changed into more productive versions. The selection was largely a hit-and-miss affair. With no clue about Mendel's factors, early farmers did little more than hope that the next generation would keep the traits that provided the most reliable food.

Over time farmers created varieties that were unrecognizable from their wild relatives. Today's corn on the cob, for example, once was a thin spikelet of erratic seeds that sprouted at the top of the plant like a roadside grass—no tidy row of plump kernels, no leafy sheaf ready for the modern barbecue. The new varieties became as domesticated as pets, transforming to suit their human cultivators. The new plants also became dependent on a care package of added fertilizer, pesticides, and insecticides all washed down with irrigation. None of them would survive today if left to grow on their own.

A mere twenty of such pampered species make up the bulk of modern agricultural production. Of those twenty, eight belong to the grass family,[1] and the most important of these is rice. Only one genus of rice, *Oryza* (from the Greek "of Oriental origin"), was chosen for cultivation, and only two species, *O. sativa* in Asia and *O. glaberrima* in West Africa.

Vegetables and fruits fared somewhat better, but as the commercial breeders began to narrow their selections, being a nice fat cabbage or a particularly juicy pear was no more a guarantee of long life than being a favorite rice plant. A 1983 survey of American publicly available fruits and vegetables showed that 97 percent of the varieties being sold by commercial U.S. seed houses had disappeared since the beginning of the century. In that period, the varieties of cabbage in the U.S. Department of Agriculture's seed storage bank dropped from 544 to 28, carrots from 287 to 21, cauliflower from 158 to 9, tomatoes from 408 to 79, cucumbers from 285 to 16, and Mendel's garden peas from 408 to 25. Of the 7,089 varieties of apple in use during the same period, 6,211 had been lost, and of 2,683 pears, 2,354 no longer existed.[2] The selection of plants for breeding from the 1960s onward reduced the number of varieties still further; a plant had to have something special to offer to survive to the end of the twentieth century.

Taipei 309, the rice type that brought the first of Potrykus's and

Beyer's golden grains into the world, had an important advantage. It was a dwarf, a tough, sturdy variety that had the ability to support a thick cluster of seeds on its short stem. Other varieties were too tall and skinny to bear the heavier yields the breeders were seeking. The taller varieties flopped or *lodged* onto the ground, where the grain rotted long before it could be harvested.

Dwarf varieties of staple crops made possible the Green Revolution, the great modernization of agriculture after World War II that by doubling world production of grain, saved millions of people in undeveloped Asian nations from starvation. One estimate suggests that breeding the dwarfing gene into wheat saved a hundred million lives.[3] The story of how plant breeders discovered the agricultural wonders of this gene began more than a century ago in the rice paddies of Taiwan.

As its botanical name suggests, *Taipei 309*'s ancestors came from Taiwan, where plants had to cope with more than farmers in their struggle to survive. Thousands of cultivated rice varieties were eliminated in the early 1900s by the Japanese who invaded Taiwan, then called Formosa, during wars with Russia and China. As a result of its empire-building activities, Japan was constantly running short of food for its troops and its growing urban workforce. It looked to Taiwan's rice paddies, whose varieties had been much admired by Japanese emperors, to make up the shortfall. To the Japanese palate, however, the common Taiwanese varieties were inferior, the most widespread type being hard and red in color. The Japanese preferred softer rice, and the whiter the better. Japan's occupation forces in Taiwan launched a ruthless campaign to eliminate the red rice and other unwanted types, cutting the number of local varieties by two-thirds, from 1,197 to 390 by 1910.

The Japanese made excellent use of the surviving varieties, however, selecting the best for their own rice-breeding program. Their lack of arable land forced them to be inventive about their choices. If they wanted to increase production of grains, they could not put more land under the plow, like the Americans (between 1866 and 1900, the total number of U.S. acres planted to corn tripled and production quadrupled) or the Russians with their vast stretches of farmland. They had to increase the yield of each plant where it stood.

To boost yields of wheat and rice, Japanese farmers pioneered the industrial agriculture so reviled by environmentalists and ecologists today. They started using commercial fertilizer—mainly organic soybean cake—which suited their new varieties but required deep plowing. This meant new plow designs and the use of draft animals instead of human labor. It also meant new irrigation systems, because the fields had to be drained before they could be plowed. Japanese farmers also introduced double-cropping—planting two separate crops a year on the same field. The more they doused their wheat and rice plants with fertilizers, the heavier the ears and panicles became, until they fell over. The farmers discovered that this lodging could be prevented if the plant were a dwarf with a shorter and sturdier stem.

As early as the beginning of the Meiji Restoration—around 1868—Japanese agriculture had changed dramatically with the introduction of dwarf varieties. By the end of the nineteenth century, the Japanese were breeding the most efficient food crops in the world; it would be several decades before the West matched their scientific advances. Some of the more famous dwarfs would come from Taiwan, having survived Japanese efforts to extinguish them. An early rice favorite was called *shinriki,* which in Japanese means "power of the gods."[4]

At first, Westerners had viewed the Japanese obsession with dwarf plants as a mere curiosity, like their miniature gardens. When Horace Capron, the U.S. Commissioner of Agriculture, visited Japan in 1873, he wrote home, "The Japanese farmers have brought the art of dwarfing to perfection." But he didn't bother to send any seeds to farmers back home. At the time, American farmers judged plants mostly by size. They were proud of their hefty varieties, showing them off at annual shows. In the Corn Belt the criteria for a fine ear of corn were to a large extent how it looked—its size, shape, color, silky tips, and all.

Like the old hunter gatherers, Americans preferred a tasty-looking crop. Big produce won the prizes in the county fairs. An ideal ear of corn was ten and a half inches in length and seven and a half inches in circumference, with nice, plump kernels that had "well-rounded butts." [5]

At the turn of the century, the Division of Botany of the U.S. Department of Agriculture decided that dwarf plants in Japan deserved a closer look. Professor Seaman Knapp of Iowa State University was dispatched to Japan. He sent home two tons of rice seed, including the famous short-grained and god-powered *shinriki*. The new varieties were planted in the Carolinas and Louisiana. *Shinriki* produced high yields but was eventually abandoned in favor of a longer-grained Honduran rice. The American obsession with size applied even to the length of the grain.

Japanese dwarf varieties would not make a decisive contribution to American agriculture until after World War II. General Douglas MacArthur's occupation force in Japan included U.S. government officials advising the Japanese on how to put their war-ravaged economy back together again. One of these officials was a U.S. Department of Agriculture employee named Cecil Salmon, who spotted the same dwarf wheat and rice seen seventy years earlier by Commissioner Capron. One in particular, a variety of wheat named *Norin 10*, caught Salmon's eye. He sent back some seeds for experimental sowing.

Norin 10, as it turned out, was part American. Japanese plant breeders had created this variety in 1917 as a cross between a local dwarf and an American wheat variety named *Fultz* after its breeder, Abraham Fultz, a Pennsylvania wheat farmer. The *Fultz* wheat was itself the product of generations of careful selection by American farmers, but only the Japanese had realized its full potential. After World War II, the dwarfing genes of *Norin 10* and other Japanese types were quickly incorporated into U.S. varieties and then into the Green Revolution. In the postwar developing world millions of lives were at risk from starvation, and in a stunning triumph for technology, the United

States would deliver the latest farming techniques and improved varieties of staple crops free of charge to undeveloped nations in Latin America and Asia.

To critics of the Green Revolution, the dwarf varieties would become a symbol of the basic flaw in the new system of modern agriculture— the creation of monocultures, or single variety, genetically uniform crops vulnerable to disease. *Taipei 309* would become a target of the antibiotech forces because it was a product of the Green Revolution and part of what one critic called "the plague of sameness."[6] That plague brought one variety of food plant to millions of acres worldwide, one type of agriculture for the developed world and the same type for the undeveloped world. Before long, it seemed, there would be only one lonely variety of cabbage and one solitary type of cucumber.

Monocultures inevitably squeezed out the local traditional varieties that had fed, housed, clothed, and cured people throughout history. Scientists invented new terms to fit the passing of so many ancient plants—*genetic erosion* and *loss of biodiversity*. Over time, these terms would become political slogans as well as descriptions of ecological phenomena.

The ancient gene pools of the lost plants held raw materials essential for making the farming revolution. To plant breeders, access to these endangered gene pools was like access to oil wells for motor cars. If there was no gene pool to provide a constant flow of new and different genes to keep a cultivated plant healthy and robust, the plant's useful life, at least to humans, was over.

The alarm about sameness had been raised early in the United States. The 1936 U.S. Department of Agriculture Yearbook warned, "In the hinterlands of Asia there were probably barley fields when man was young. The progenies of these fields with all their surviving variations constitute the world's priceless reservoir of germ plasm. It has waited through long centuries. Unfortunately, from the breeder's standpoint, it is now imperiled. . . . When new barleys replace those grown by the farmers of Ethiopia or Tibet, the world will have lost something irreplaceable."[7]

The real fear, however, was the loss of these gene pools in the early centers of human civilization where the staples had been cultivated—the Middle East, India, Southeast Asia, Mexico, and Peru, the so-called centers of diversity of food plants. In those dozen or so centers, the cultivation of plants by early farmers had produced thousands of different varieties, but none of the key centers was in the developed world. Plant breeders were constantly launching expeditions to tropical countries to discover one more exotic gene.

In exchange for giving away their technology, American seed companies would gain access through the Green Revolution to these invaluable gene pools. The U.S. program's humanitarian goal was the conquest of hunger, but as University of Wisconsin rural sociologist Jack Kloppenberg commented, the overall strategy for spreading the Green Revolution was a "volatile mix of business, philanthropy, science, and politics."[8]

A green revolution today suggests a radical environmental movement, but this one was nothing of the kind. Toward the end of World War II, the United States decided to use its dominance in world food production to extend its global influence. Food would become a political device. In a dual strategy the United States would simultaneously fight world hunger and halt the spread of communism. Population explosions in the undeveloped nations of Latin America and Asia meant there was not enough to eat. Hunger led to social upheaval that could leave nations vulnerable to communist takeover. The Roosevelt administration was especially concerned about political stability on America's southern border. Mexican agriculture was in crisis, and the United States feared a peasant uprising.

At the time there was no United Nations or coordinated international aid program to help countries improve the yield of their crops. The U.S. administration launched its Good Neighbor policy, promoting U.S. interests without military intervention. Two of America's leading philanthropic foundations, Rockefeller and Ford, put together an emergency crop improvement program, first in Mexico and then in the rest of Latin America, in India in the 1960s, and in other developing nations in Asia in the 1970s.[9]

When it was over, an enterprising U.S. government bureaucrat from the Agency for International Development would describe this boom in staple crops as the "Green Revolution." *Green* referred to swaths of young green shoots of corn, wheat, and rice that suddenly took hold in lands that had previously produced sparse harvests. *Revolution* referred not to upheaval of the masses but to the combined effect of improved seeds, chemical fertilizers and pesticides, and water irrigation projects.[10] The true birthplace of the revolution, however, was not Mexico or Colombia, the Philippines or Surinam, but the experimental farms of the advanced industrialized nations, first Japan, then the United States and Europe.

In America the modern farming revolution really began with the development of hybrid corn. American farmers in the 1800s crossbred corn using varieties cultivated over centuries by Native Americans. Around 1870, breeders began to notice the phenomenon of *hybrid vigor*. When one variety was crossed with another, the offspring were generally in better health. Bigger and stronger than their parent plants, the new varieties produced more seeds—and more food.

After 1935 hybrid breeding became more sophisticated. The latest hybrids were created by an artificial cross between two varieties that had been inbred, fertilized by their own pollen for three or four generations. The first generation of crossbred plants showed a tremendous leap in hybrid vigor, with grain yields up to 50 percent higher. How the plants come to produce such energy is still not fully understood. Some geneticists doubt the entire concept.[11]

Whatever the scientific reasons for the extra vigor, the phenomenon presented corn seed companies with a business opportunity that gave the color concept of green "green" a new meaning. Hybrid vigor only lasted for one generation. Once the hybrid was openly and naturally fertilized in a farmer's field, the gene pool was again mixed up and the vigor of the hybrid progressively declined in successive generations. If farmers wanted to maintain the much bigger yields from hybrids, they had to buy seeds every year.

The hybrid seed industry flourished. An early promoter of corn hybrids was Henry A. Wallace, plant breeder and founder of the Hi-Bred Corn Company, the first company devoted specifically to the commercialization of hybrid corn and the forerunner of Pioneer Hi-Bred International, which would become America's largest seed company. Wallace is famous for his cheerleading on hybrids, especially the remark, "We hear a great deal these days about atomic energy. Yet I am convinced that historians will rank the harnessing of hybrid power as equally significant." In the decade 1934–44, hybrid seed corn sales leapt from virtually nothing to more than $70 million. Seed corn became the lifeblood of a new vibrant seed industry. As Jack Kloppenberg would observe, "Hybridization [was] a mechanism for circumventing the biological barrier that the seed had presented to the penetration of private enterprise."[12] Between 1950 and 1980, hybrid seed corn sales went up 60 percent. The period 1950–70 saw the virtual disappearance of the farmer as a producer of his own seed corn.

In 1969 only six hybrids accounted for 71 percent of corn grown across America. The new industrial agriculture had arrived, bringing corn monocultures to millions of acres. But then the early warnings of the plant breeders about genetic uniformity suddenly became a reality. In 1970, 15 percent of the U.S. corn crop was lost to southern corn leaf blight at a cost of $1 billion. In some states half of the corn crop withered in the field. One variety used in 85 percent of the hybrid corn plants was implicated. Corn plants across the country had become "as alike as identical twins," the National Academy of Sciences reported. "Whatever made one plant susceptible made them all susceptible." The academy concluded ominously that U.S. agriculture was "impressively uniform genetically and impressively vulnerable."[13]

But there was another side to the hybrid story. The new varieties had encouraged not only the seed companies to enter the market but a whole international commercial enterprise that would become known by a new generic term, *agribusiness*. In addition to the companies that produced the new improved seeds were companies that made the chemical fertilizers, insecticides, and pesticides on which the seeds' superior performance depended. To extract the maximum yield, hybrids required an expensive chemical mix.

World War II had created a huge production capacity for nitrogen, the key ingredient of military explosives but also the main ingredient of plant fertilizer. When the war was over, factories that had produced explosives were converted; mineral fertilizer production worldwide rose from 17 million tons to today's figure of more than 150 million tons.[14] Over the three decades from 1950 to 1980, the sale of nitrogen fertilizer jumped seventeen times.[15]

To make the most of the result, along came machinery to plant and harvest the corn. Tractors replaced draft animals and their housing, feed, and veterinary costs. In 1938 only 15 percent of American corn acreage was harvested by machine, and for good reason. Different varieties of corn grew different numbers of ears at different rates and at different places on the stalk and one machine doesn't handle all the different varieties. Hybrid breeders were able to shape the plant to the machine; by 1945 the amount of corn mechanically harvested had jumped from the original 15 to 70 percent. Between 1930 and 1950, the number of mechanical harvesters increased ninefold.[16]

The Rockefeller plan was to transfer this new farming technology to undeveloped nations, but U.S. corn and wheat seeds did not grow well in Mexican fields and even had difficulty when crossbred with Mexican varieties. The climate was different, the soil was poorer, and Mexico's principal corn-growing areas were at high elevations, up to seven thousand feet. In an initial test, all the U.S. corn hybrids failed. By 1948, however, five years after the program was launched, hybrids were flourishing and for the first time since the Revolution of 1910, Mexico had no need to import corn. A year later the best new varieties were yielding nearly 50 percent more grain. By the 1960s more of Mexico's corn lands were planted with hybrids, and total production had gone from two to six million tons.[17]

Producing enough food to stay ahead of the increase in world population was the daunting mission of a small number of dedicated plant breeders around the world. The most famous American was Norman Borlaug, a Midwesterner and plant pathologist who oversaw the Rockefeller Foundation's wheat program in those early years of the Green Revolution. Borlaug's first task was to breed wheat varieties that were resistant to disease, especially the debilitating wheat stem rust. Mexico's wheats had lost all their defenses against wheat rust. But Borlaug's big success was the introduction of the dwarfing gene from the Japanese *Norin 10* into the Mexican wheat varieties.

Borlaug began increasing wheat yields by dousing the Mexican wheat plants with nitrogen fertilizer. The problem was that the higher-yielding plants were so heavy they fell over before they could be harvested—just as their northern cousins had done before U.S. breeders had embraced the dwarfing gene. Borlaug sent for the *Norin 10* dwarf seed and the problem was solved. The new wheat plants were twenty to forty inches tall, compared with the fifty to sixty inches for the traditional varieties.[18] Mexican wheat production soared.

All Mexicans benefited in some way from the sudden rise in food grain production, but the new package—seed, fertilizer, and other chemicals needed to produce the new yields—was too expensive for all but the richest farmers. A wealthy group of fewer than two hundred millionaire entrepreneurs established themselves quickly in the fore-front of Mexico's food production.[19] The Mexican government en-couraged this elite. The state offered sources of credit for farming operations and private irrigation schemes, fostered mechanization by special exchange rates, and established guaranteed prices for wheat. By 1951 Borlaug's new wheat varieties were being grown on 70 percent of the wheat fields. Five years later Mexico was self-sufficient in wheat. At the same time, wheat yields of the poorer farmers dragged behind those of the more prosperous commercial sector.[20]

As a result of helping to set up a Mexican agribusiness, U.S. seed companies gained long-term access to Mexico's priceless gene pool of traditional corn varieties. As part of the Mexican program, the Rocke-feller Foundation opened the International Maize and Wheat Im-provement Center, a research center and collection of genetic information on native Mexican corn, including a seed bank that would eventually preserve more than one hundred thousand varieties. The seed companies were happy, but the primary purpose of these seed banks was to hold their valuable collections in trust for all hu-manity.

Similar publicly funded agricultural research centers and associated seed banks were set up in the Philippines for rice, in Colombia for beans, and eventually in Africa for rice, cassava, and maize. By the early 1990s sixteen such centers were spread over five continents, but the African centers still lagged behind the rest of the world because farmers there could derive little benefit from the technical package of-fered by the West. In some cases the crop was inappropriate, in others the African farmers simply could not afford the chemicals being of-fered. The research laboratories were all coordinated by the Consulta-tive Group on International Agricultural Research (CGIAR), research centers financed by the advanced industrial nations, private founda-tions and international regional organizations, the Rockefeller and Ford foundations, and the World Bank. One aim was agricultural, the

other geopolitical. "The Green Revolution operated in areas susceptible to communism," concluded a Dutch study.[21] In some accounts,
this aspect would obscure the humanitarian mission.

After Mexico, Borlaug and the Rockefeller and Ford–sponsored research teams scored similar successes with wheat and rice programs in
Asia, where population was increasing at an alarming rate and traditional farming methods simply could not cope. In the early 1960s the
wheat yield in India and China was similar to that of Europe during
the Middle Ages. Famines followed if crops failed. The expansion of
farm acreage was reaching the limit of arable land.

The Green Revolutionaries introduced the same three-part package of new varieties, irrigation systems, and chemical fertilizers. Production soared. The 1968 wheat harvest in India was one-third greater
than the previous record; schools had to be closed to provide space to
store the grain. One study showed that the increase in India's wheat
production from the new varieties was so great, another 100 million
acres of land would have had to be plowed up to obtain the same yields
with old varieties. Worldwide, between 750 and 1,200 million acres
would have been needed, according to the study.[22] The new wheat varieties were also popular in Latin American countries.

Again the dwarfing gene was decisive. After the new rice research
station started distributing a dwarf rice in the mid-1960s, the Philippines became self-sufficient in rice production for the first time in decades. In Colombia the new rice did so well, it became the country's
dominant food crop. In 1970 Borlaug's labors were rewarded with the
Nobel Peace Prize; he was the first plant breeder to receive the honor.[23]

Geopolitics apart, an increase in food production was the top priority of these programs, and as a result they saved millions of people
from starvation. The statistics are stunning. The total amount of food
available per person in the world rose by 11 percent over the two decades of the '70s and '80s, while the estimated number of hungry people fell from 942 million to 786 million, a 16 percent drop.[24]

In the early 1980s "the all clear was sounded, signifying a job well done," as the director of the International Food Policy Research Institute (part of the CGIAR consortium), Per Pinstrup-Andersen, would put it. But he and others were worried about the fallout from this success. M. S. Swaminathan, the international agronomist and promoter of the Indian Green Revolution, would recall how in the same year, 1968, that the Green Revolution had been given its name, he had warned, "Exploitative agriculture offers great possibilities if carried out in a scientific way, but poses great dangers if carried out with only an immediate profit motive. The emerging exploitative farming community in India should become aware of this. Intensive cultivation of the land without conservation of soil fertility . . . would lead, ultimately, to the springing up of deserts. . . . Therefore the initiation of exploitative agriculture without a proper understanding of the various consequences of every one of the changes introduced into traditional agriculture, and without first building up a proper scientific and training base to sustain it, may only lead us, in the long run, into an era of agricultural disaster rather than one of agricultural prosperity." In 1999 Swaminathan observed, "The significance of my 1968 analysis has been widely realized." [25]

The Green Revolution had not brought world hunger under control. Even the doubling of food production would still leave an estimated 800 million "food insecure" in 2025. And while the successes had averted famine for millions and India's granaries might have been overflowing, five thousand children died in India each day from malnutrition. While increases in the yields of three staple crops—corn, wheat, and rice—helped the poor raise their living standards where those crops are eaten, improvement in African staples, such as cassava, sorghum, and sweet potatoes, received lower priority and became known as "orphan crops."

The worst of the unintended consequences of the Green Revolution was the damage caused by agricultural chemicals and bad drainage.

The more luxuriant plant growth that resulted from use of the chemicals provided ideal conditions for insect, disease, and weed buildup that in turn encouraged the use of insecticides, fungicides, and herbicides. Heavy use of insecticides and pesticides poisoned more than insects and pests. Rachel Carson's *Silent Spring* warned that these chemicals in developed countries were beginning to cause more than high yields. Dramatic declines had been seen in bird life; their silence had become an alarm bell. As the Green Revolution moved east in the late sixties and seventies, pesticide use soared; Asia reached one-fifth of world consumption by the mid-1980s. In India chemically treated areas expanded from 15 million acres in 1960 to more than 200 million acres in the mid-1980s.[26] There were hundreds of thousands of accidental human poisonings—from inadequate supervision, poor labeling, and an unwillingness to wear protective clothing—and several thousand deaths. Pest populations grew in response to higher nitrogen applications. Plant diseases increased in the microclimate created by the densely leaved, short-strawed wheat and rice plants. Many pests began to build up resistance to the chemicals. As a general rule, the more they were sprayed, the more quickly they became resistant.

The chemicals killed not only the pests but also their natural enemies. One of the worst cases concerned the brown planthopper, a sucking bug that attacks rice plants. It was almost unknown in Asia before the introduction of the new rice varieties, but it caused devastating losses in Indonesia, in particular, in the 1970s and 1980s. Scientists working in Asian rice fields discovered that the planthopper is normally controlled by parasites that destroy its eggs and by the wolf spider that preys on the planthopper itself.

These natural predators were gone, destroyed by the new agricultural chemicals. Researchers began to talk about the "pesticide treadmill." In some areas the population of the planthopper rose in direct proportion to the amount of insecticide sprayed.[27] One answer was to breed new varieties that would include a gene that made rice plants resistant to the planthopper. This proved successful, but only for a short period before the pests adapted to the crop's new defenses. A second answer was to produce plants with a combination of genes that provided only a partial defense against the planthopper, but which cut back the need for pesticide use. Some plants were lost to the pest, but the planthopper's ability to adapt was slowed. However, the most successful response was an ecological one—when farmers used the minimum amount of pesticide that also allowed natural predators to help in destroying the invaders. Using this "integrated" approach, overall yields rose by 15 percent while pesticide use declined by 60 percent.[28]

In the end the Green Revolution was more a triumph of American technology than of science. In the United States, farmers who survived were either big enough to keep expanding their acreage or small enough not to enter the industrialized agriculture race at all. The first group became the superfarms, with half a million dollars or more in annual sales. The second, much smaller group, became the organic farmers. The rest dropped out. The death of family farms became an all-American saga. In the United States the number of farms has been cut by two-thirds since World War II, and the average farm size has more than doubled.

This spiral of technological advance and social disruption of the farming community could also be found when the Green Revolution programs took effect in Latin America and Asia. In Mexico the technology only worked for certain farmers who could afford the inputs of chemical fertilizer and pesticides. The disparities between rich and poor farmers were also found in India and other parts of Asia. In Bangladesh the new variety package cost 60 percent more than the package for traditional varieties. In the Philippines, where the majority of farmers are tenants, landlords loaned cash for improvements at rates of 60 to 90 percent a year, "often producing a permanent state of indebtedness."[29]

To those who campaign today against genetically modified foods, the Green Revolution is an agricultural cautionary tale, a vivid lesson not only on the specific dangers of monocultures but also on the way world farming has changed from a publicly financed, farmer-driven occupation that allowed even the poorest peasant to take part into a raw industrial enterprise, the mass production not of machines and widgets but of food.

In many ways the social disruptions of the Green Revolution were inevitable; the history of agriculture has provided many examples of how new technologies are never neutral but are rather associated always with major shifts in society. Consider the colonies of the European empires in the eighteenth and nineteenth centuries. Farming in the tropics concentrated on cash crops such as tea, cocoa, sugarcane, pineapples, dates, bananas, and spices, not the crops needed by indigenous peoples. The early agribusiness of the empires included "a whole suburban world of brokers, shippers, underwriters, warehouse clerks, bankers, retailers, and rentiers [who] lived prosperously on the differential between the cost of the production of these cash crops in the fields of Asia, Africa, the Caribbean and South America, and the selling prices in the metropoli," as Andrew Pearse wrote in *Seeds of Plenty, Seeds of Want,* about the social effects of the Green Revolution.[30]

Large-scale farmers adapted to the new package more rapidly than smaller farmers, who couldn't afford it. But in the view of the food and population expert Per Pinstrup-Andersen and his coauthor Ebbe Schiøler, "This outcome was unavoidable. Even though small-scale farmers would like to produce more and earn more, their first concern must be to avoid loss, so they seldom dare to risk everything on a new technique or variety. But once they see the results achieved on larger farms, they become as actively involved as the large-scale farmers."[31] The vicious cycle of overproduction and declining farm prices was under way.

There was one disturbing aspect of the Green Revolution that everyone acknowledged, however. The magnificent increases in yields of wheat and rice seen in developing countries in the early years between 1967 and 1982 had begun to decline below the potential projected by the breeders. In the Philippines, for example, rice yields grew steadily during the 1970s, peaked in the early '80s, and the growth rate has dropped off gradually ever since. Similar patterns were seen for rice and wheat in India and Nepal.

In some cases the decline was expected. As the Green Revolution spread, the new varieties were planted in poorer soils by poorer farmers who did not have the resources to add sufficient nutrients to boost the plants to their full potential yield. However, researchers also began to notice declines in the yields of plants in the best soils. In Thailand and the Indian Punjab, farmers spoke of soils that had become "lifeless," with severe declines in certain soil nutrients, such as zinc, and an increase of toxic chemicals from irrigation wells.

The growing scarcity of water became an urgent concern as overuse of irrigation wells resulted in the appearance of harmful salts in some soils. Soil erosion also caused a drop in yield or made it impossible to continue to sow crops, particularly in very hilly country. The depletion of nutrients in soils was especially worrisome in Africa, where each farmer typically has little land and can afford only insufficient amounts of fertilizer to replace nutrients. Critics spoke of the Green Revolution as a mining operation—extracting maximum output from the land in the shortest possible time.[32] Such declines in the quality of soils and the quantity of water were especially distressing to population experts who predicted that the world's farmers would have to harvest 40 percent more grain in 2020 than they did in 1995.

Some saw the future in genetic engineering, in the insertion of alien genes into food plants to create crops that could cope with the salty soils and better extract whatever nutrients were available. Perhaps the key to increasing food production in developing countries would be found in a mix of traditional breeding backed by genetic modification, a "doubly green revolution."[33] But as always in this biotech era, more was at stake than agricultural research and improved plant varieties. Before the promises of gene splicing could be realized, a wary public had to be persuaded that the genes being transferred into the new modified plants were not making the new foods unsafe.

A New Sort of Tomato

If a new food, or food component, is found to be substantially
equivalent to an existing food or food component, it can be
treated in the same manner with respect to safety.
—OECD, 1993

Because a GM food is "substantially equivalent" to a conventional one does not mean that it is "safe."
—British government report, 1999

In the early days of bioengineering, two flowering plants were particularly hospitable to alien genes. One was tobacco, which is perhaps the
easiest plant in the world for biotech experiments. The other was the
modest but cooperative petunia. A petunia field test in Germany
caused an unusual stir in the biotech community during the summer
of 1990.

Researchers planted thirty-one thousand white petunias into
which they had inserted a gene that they thought would turn the flowers red. The first flowers were indeed red, but in the next batch, the
color had begun to fade. By midsummer the flowers were all white,
and by the end of the season the flowers were a mixture of red and
white. The strange variations could not be explained by Mendel's laws
of heredity. It turned out that a midsummer heat wave had simmered
part of the gene that was supposed to help the red gene do its work.

Such unpredictable genetic transformations of the new technology prompted the British biology writer Colin Tudge to suggest that it would be more appropriate to refer to genetic engineering as "genetic gardening," a glorious image of the colorful surprises of an English herbaceous border, but not quite the picture that the biotech companies want to portray.

They prefer the term *engineering* because they want to project the idea of assembly-line precision. To calm worried consumers, the companies like to say that genetic engineering is really only an extension of the traditional art of crossbreeding, as practiced by Mendel in his monastery garden. The modern advantage, the companies argue, is that genetic engineering is more fastidious and targeted, thus the results are more predictable and, by implication, the product is safer.

Opponents of biotechnology hotly dispute this assertion. They accuse the companies of distorting the realities of traditional breeding and turning a blind eye to the possibilities for genetic, environmental, or human harm that could result from inserting an alien gene into a plant's genome.

The companies' version goes something like this: When Mendel dusted pollen from the flowers of a wrinkled pea onto the flowers of a smooth pea to breed a new variety, he did not know precisely what other changes he might be transferring in the next generation of peas. A plant contains tens of thousands of genes, and Mendel transferred the gene for wrinkling along with thousands of others for different traits—such as height, seed color, or petal location. No traditional plant breeder has control over these so-called hitchhiking genes. By contrast, bioengineers select a single gene for a single characteristic and insert the gene into a plant where, if everything works properly, the new trait appears.

In this narrow, theoretical sense, bioengineering is indeed more precise, but the theory misses much of what's going on when a gene from one species is transferred into a plant of another species. It ignores the possibility that something can go badly wrong.

Opponents of the technology argue that in traditional breeding, each plant is of the same or a related species and the mixing or recombination of genetic material occurs between two plants that share a recent evolutionary history. Traditional plant breeders, the argument goes on, generally know where the gene or genes they want to transfer are going to end up on the host plant's genome. The result, usually, is no major disruption in the plant's inner workings, and most offspring of traditional breeding are normal. The seeds germinate and produce plants that themselves go on to reproduce without creating radically different or defective progeny.

At this stage in genetic engineering, the opponents continue, transferring the new gene is a very inefficient process, with only a tiny percentage actually arriving on target. Among the failures are unstable plants that fail to reproduce the inserted trait in successive generations.

Part of the problem is that the new gene for a desired characteristic is not inserted into a plant's cell on its own, as the company imagery suggests, but as part of a package of genes. The package includes the complex signaling system of DNA promoters that control where and when the new gene is switched on in the transgenic plant. It also has a stop signal to switch the new gene off and marker genes to show scientists the plant cells that have successfully received the package.

The gene package is often referred to as a cassette because it contains information that is fed into the transgenic plant as you might feed a tape into a tape recorder. There is a big difference, of course. Your tape recorder will then play what is on the tape. But genetic engineers can't control what the gene cassette will "play"—because genetic engineers have no real control over precisely where the cassette lands in the chromosome of the host plant. The cassette could land in the middle of another gene's DNA, where it could give or receive different instructions from those intended. With communications gone haywire, the new gene for the new trait might never be promoted, or read, by the host plant's cells, and the desired protein might never be produced.

Another worry, say opponents, is that the gene might produce too much of the new protein, turning the plant into something that is either useless or, worse, toxic. Plants are complex biological systems; many of them, the tomato for example, have some parts that are tasty and safe to eat and other parts that are poisonous, such as the leaves and the stems. If alien genes land in the wrong place, they might start to produce these toxins in the edible parts of the plant. The new gene cassette could also trigger a reaction that produces an allergen, a substance that causes an allergic reaction in humans or animals.[1]

Although such worries were only theoretical, people allergic to peanuts or any kind of nut, for example, had special cause for alarm that once-safe foods would suddenly become toxic without their knowledge. In the early 1990s word leaked from the biotech laboratories that researchers were looking at proteins in nuts with the idea of isolating the gene that made them so nutritious and then transferring it into a major crop, such as soybeans. (Their goal was a lofty one, similar to Potrykus's and Beyer's efforts with golden rice.) They found a gene in the Brazil nut responsible for making certain amino acids—nutrients the soybean lacks. Pioneer Hi-Bred International, the biggest seed company in the United States, tested the gene to see if it caused an allergic reaction. It did, so Pioneer dropped the project and published the results. But somehow the story lived on. In some versions, the allergen in the Brazil nut had been identified too late to stop it from being incorporated into a soybean crop.

In short, the practical application of genetic engineering at an early stage was no more predictable than that of any other new technology, from the steam engine to nuclear power. Just as the engineers of nuclear power had downplayed the risks of cooling nuclear reactors, so the biotech companies minimized the scientific realities of making a transgenic plant. In the same breath, companies were stressing the novelty of bioengineering techniques to their investors while assuring the public that these changes in the food supply presented no risks to health or the environment.

The petunia and the Brazil nut experiments, as harmless as they were in the end because they were not developed into foods, sent mixed signals to wary consumers. They showed how unpredictable the technology could be and gave warning that the biotech companies were concealing details about this business, including some that might be harmful. While genetic engineers agreed that unintended and unexpected results can and do happen, they insisted that any harmful strains could be eliminated well before a food reaches the marketplace. In any case, they insisted, virtually every plant cross, whether performed by nature or plant breeders, can also result in major and unexpected rearrangements of the genome. Plants have very plastic genomes. Finally, the engineers claimed, the genetic instability complained of by opponents is not really a health or safety issue for the new foods, but a problem that can and would be worked out in the laboratory.[2]

At the beginning of the 1990s, as the first GM foods were ready for market and the bright future of the new technology was being promoted on Wall Street, these scientific opinions had to be translated by governments into safety regulations for the new foods.

Traditionally, government regulators compare new foods, or any consumer product, with the closest previous version. Either the new products are regarded as generally the same, and therefore do not require any new regulation, or they are different enough to be tested separately before being declared safe. Politically, the Reagan-Bush administration, which promoted government deregulation, was inclined to give U.S. companies every opportunity to exploit the new technology without being weighed down by burdensome rules.

Despite persuasive scientific arguments that biotech engineering was more than just an extension of traditional breeding, and despite a general acknowledgment that the nation's laws governing food safety were not equipped to take care of the new science, the administration decided to treat the new foods as no different from the old ones. It was the product that mattered, not the process. As long as the company producing the new food assured the government that its product was free of new toxins and allergens, it was to be "generally regarded as safe," or GRAS, in the acronym of the government's regulator, the Food and Drug Administration. The thinking was, if it looks like a tomato, tastes like a tomato, and the company promises that it is very similar to the old tomato, then it is still a tomato.

Advice from the National Academy of Sciences, the foremost scientific body advising the government, supported the administration's view. The academy declared that there was "no evidence, based on laboratory observations indicat[ing] that unique hazards attend the transfer of genes between unrelated organisms. Furthermore, there is no evidence that a gene will convert a benign organism to a hazardous one simply because the gene came from an unrelated species."[3]

The Reagan administration assumed that the biotech companies would be relieved that there were no new rules, but Monsanto, the leader in agricultural biotechnology, was worried. The company's executives called on Vice-President George Bush and "bugged him" for better regulation."[4] Monsanto had looked into the future acceptance of GM foods and recognized earlier than most that while the use of gene splicing in the development of lifesaving drugs, such as insulin, was unlikely to draw fire, "monkeying around with plants and food would be greeted with deep skepticism."[5] The company knew that there were some genuine scientific concerns about the process of genetic engineering. They also knew that the antibiotech groups were preparing for an all-out war. Monsanto, the "wicked old chemical giant," was a prime target.

Leading the opposition in the United States in those days was Jeremy Rifkin, the veteran liberal campaigner, who became the loudest and one of the most effective foes of biotechnology. He had formed a broad coalition that included environmentalists, creationists, and family farmers, all vowing to stop any kind of biotechnology, not only gene splicing in plants but also in animals. His weapon was the lawsuit, his political ally a young senator named Al Gore. Monsanto had watched Rifkin in action and recognized trouble. The company was investing millions of dollars to research transgenic plants for agriculture. They wanted a smooth launch of their products, and that depended on their winning the public trust.

Monsanto executives began urging the Reagan administration to come up with some kind of rules, not too rigorous of course, but some government requirement or procedure to help a jittery public feel confident that these new foods had been endorsed by the government as safe to eat. "The biotechnology companies wanted government regulators to help persuade consumers that their products were safe, yet they also wanted the regulatory hurdles to be set as low as possible," researchers at the University of Sussex in England observed.[6] In 1990 the first Bush administration issued a set of regulatory "principles" that concentrated only on the product and urged federal agencies to conduct speedy reviews of new foods to "minimize the regulatory burden."[7]

The FDA Commissioner in 1990 was an earnest young pediatrician named David Kessler, who would become nationally famous for flexing his regulatory muscles against the tobacco companies. Kessler, who was still finding his way through the agency's arcane bureaucracy, gave his full support to the Reagan-Bush stance, overruling scientists in his office who warned that genetic engineering was unknown territory that might create unexpected risks in plants and even people. FDA scientists knew perfectly well that novel toxins might be produced when alien genes were inserted into a plant's genome. They could not, as scientists, support the opposite assertion that there could be no unintended effects. In fact, the FDA scientists argued, there was ample scientific justification to require tests for each new modified food.[8]

In an internal memorandum, Linda Kahl, an FDA compliance officer, argued that the differences between traditional plant breeding and genetic engineering led to different risks, but there was no data that addressed the relative magnitude of the risk. Another FDA scientist, Louis Pribyl, added that the agency's insistence on regulating the product and not the process meant that the agency's proposals "read very pro-industry, especially in the area of unintended effects." It was the "industry's pet idea," he said, that "there are no unintended effects that will raise the level of the FDA's concern."

In another internal FDA memorandum, Kessler considered that it was "critical" not only to provide the biotech companies "with a predictable guide to government oversight, but also to help them win public acceptance of these new products." He believed in their creations. "The new technologies give producers powerful, precise tools to introduce improved traits in food crops, opening the door to improvements in foods that will benefit food growers, processors, and consumers," he wrote.[9]

Kessler warned, however, that a coalition of green groups opposing the proposed regulations was gaining strength. The greens wanted the new foods treated as though they contained a food additive, meaning that each new product would have to undergo a separate test, causing extra expenses and delays. They also argued that consumers deserved to know what they were getting and that the new product should be labeled. The prospect of required labeling was a nightmare to food processors trying to protect their brands from anything that might raise consumer worries about GM foods. Kessler correctly predicted that the opposition "may challenge our policy as leaving too much decision making in the hands of industry and not adequately informing customers."

The final Bush policy document in 1992 stated that there would be no new rules unless substantial changes were made to the nutritional composition of the foods and no need for labels unless known allergens, such as peanuts, were in the product.[10] The new rules, by promoting the idea that the new foods were "substantially equivalent" to the old ones, were designed to be an official government declaration that the foods were substantially safe. There was no mention of the way transgenic plants might produce harmful substances in an unpredictable way. And there was no reason such events should be mentioned, argued the probiotech forces. Nontransgenic plants can also produce such substances. But the policy strictly limited the regulatory reach of the FDA.[11] It was a clever idea for promoting the industry, but in the long run, it would not serve either the public or the companies well.

The new doctrine of substantial equivalence received its international recognition in 1993 from the Organization for Economic Cooperation and Development (OECD), a conference for a group of advanced industrial nations formed to develop economic and social policy. In particular, the OECD studied how countries should evaluate the safety of GM foods. Three years later the UN Food and Agricultural Organization and the World Health Organization would endorse the doctrine as well. But "substantial equivalence" was an intentionally ambiguous term, and even when these agencies tried to define it, their definitions were loose enough to allow plenty of leeway to the producers of genetic crops and foods. The OECD concluded, "If a new food or food component is found to be substantially equivalent to an existing food or food component, it can be treated in the same manner with respect to safety."

The scientific community had no real definition of the term either. British researchers at the University of Sussex argued that there was intentional fuzziness here: "The degree of difference between a natural food and its GM alternative before its 'substance' ceases to be acceptably 'equivalent' is not defined anywhere, nor has an exact definition been agreed by legislators." Then, "It is exactly this vagueness which makes the concept useful to industry but unacceptable to the consumer."[12]

The lack of a definition began to cause confusion in another direction as well. To many scientists, it was unclear whether the new foods were required to undergo extra tests or really were exempt, as the companies had hoped. A Canadian Royal Society report described the problem. "To say that the new food is 'substantially equivalent' is to say that 'on its face' it is equivalent (i.e., it looks like a duck and it quacks like a duck; therefore we assume that it must be a duck—or at least we will treat it as a duck). Because 'on its face' the new food appears equivalent, there is no need to subject it to a full risk assessment to confirm [the] assumption."

The OECD conclusion, the Canadians said, could also be interpreted as a requirement to establish scientifically "that the new food is identical in its health and environmental impacts to its conventional counterpart" and that it does not differ in any way other than the presence of a single new gene. Only after this finding is made can the new food be treated as safe. Importantly, European health officials concluded that for a GM food to be substantially equivalent to a conventional one does not mean that it is "safe." [13]

But within the United States and Canada, the confusion still worked in the companies' favor. Substantially equivalent would mean substantially equivalent enough to pass a consumer's squeeze test, little more. [14] For those who were watching what was going on in the world of futuristic agriculture, the concept provided "an excuse for not requiring biochemical and toxicological tests," as one dissenting scientist put it years later. [15]

Opponents zeroed in on the word *substantial*. They argued that the term does not take into account the subtle changes that might take place during genetic engineering and that it represented little more than a subjective judgment by the manufacturers. It was a policy, they said, that encouraged GM foods rather than scientific principles. In the view of the British researchers, the doctrine was "a pseudoscientific concept because it is a commercial and political judgment masquerading as if it were scientific." However, in the United States substantial equivalence would remain the industry and government standard, even though it demonstrably failed to attract the kind of trust from the public the companies were originally looking for. [16] Many ordinary consumers began to wonder if the antibiotech forces were right; perhaps there were, in fact, important differences between these new superfoods and their less perfect precursors.

The critics wanted regulatory agencies to take a more cautious approach by refusing to allow a product into the marketplace until a body of scientific evidence suggested that the new food would not cause any problems. In the language of government regulations this is known as the precautionary principle. Like "substantial equivalence," "precautionary principle" is loosely defined, but its emphasis is far different. The principle requires that governments err on the side of caution in the absence of scientific proof of the nature and severity of the risks a new product might pose to health and the environment. In other words, better safe than sorry.

The policy of substantial equivalence would set the United States on a collision path with European nations, with whom the precautionary principle would find favor. U.S. agricultural exports of GM crops would be seriously affected, and eventually the biotech industry and U.S. farmers would be forced to rethink their methods of assuring crop and food safety.

But in the early 1990s, long before the protests reached gale force, Monsanto had managed to get exactly what it wanted: a set of regulations with no teeth. To cap the victory, responsibility for oversight of transgenic plants and their products would be divided among three agencies. The Food and Drug Administration would oversee food safety. The Environmental Protection Agency (EPA) would look at plants engineered to make their own pesticides. And the U.S. Department of Agriculture would be responsible for making sure that new genes from the transgenic plants did not escape into farmland to harm or dilute the gene pool of traditional agricultural crops.

By the mid-1990s the policy of substantial equivalence would become a key weapon for the antibiotech forces seeking to undermine the technology. For Jeremy Rifkin, the adoption of substantial equivalence was the turning point in his personal crusade. He launched a "pure food campaign" and called for a global moratorium on biotechnology. He alerted green groups around the world that the United States was about to dump transgenic foods into the world market, new foods that were untested and unlabeled and perhaps, Rifkin hinted, unsafe. Millions of dollars started to flow from U.S. foundations that usually supported environmental causes into antibiotech groups, but the U.S. government soldiered on, adhering to the substantial equivalence doctrine throughout the Clinton administration, dismissing critics of its policy as irrational, politically motivated, and scientifically perverse. The first real test of the doctrine was the very public appearance of a new tomato genetically modified not to rot like its older cousins. Enter the Flavr Savr tomato, marvel of modern technology.

Long before genetic engineering, the American tomato had suffered several "improvements" at the hands of growers as they tried to meet the voracious American appetite for fresh tomatoes, estimated at $4 billion annually. When growers introduced mechanical pickers, the tomato had to be tough enough to survive the unfeeling fingers of the harvesting machines. So breeders had been carefully selecting varieties that ripened more slowly, or ones that had tougher skins. Still, all that breeding made little difference to the tomato's ability to survive the picker until growers introduced ethylene to their harvesting process. Ethylene, a gas once used as an anesthetic, is also a ripening agent produced in tomatoes, bananas, apples, pears, and most stone fruit.

Growers could now harvest tomatoes when they were green, hard, and immature (which solved the mechanical picking problem), ship them to market, and then treat them to a whiff of ethylene. The green fruit quickly turned red, or at least pink enough to go into the grocery bins. The customer never knew what had happened—except that the tomatoes no longer tasted like tomatoes. But Americans' complaints were muted, mostly because they continued to eat their fresh tomatoes in sandwiches stuffed with bacon and other fillings, or hamburgers doused with so many toppings that the taste of the tomato was secondary to its crunch or color.

The rot-free tomato came to life in the late 1980s when a group of bright young biotech researchers at Calgene, a start-up company in Davis, California, wondered what would happen if the gene that starts the rotting process could be delayed. The tomato would ideally turn red on the vine yet remain tough enough to be mechanically picked. Such a tomato could be marketed with an alluring label that promised "vine-ripened" fruit. At the very least, the researchers figured, a tomato that went bright red on the vine had to taste better than one that had been gassed from green to pink. Scientists in genetic engineering labs around the world in the 1980s were obsessed with such clever gene tricks, but Calgene thought this one had real promise.

The Campbell Soup Company thought so too and funded the research. Calgene researchers soon discovered the rotting process was controlled by an enzyme called polygalacturonase, or PG for short. PG broke down the soft inner flesh of the tomato. The researchers found the gene that produced PG and simply reversed its DNA sequence, a genetic engineering trick known as *antisensing*. The first crop of modified tomatoes turned red on the vine. They were tough enough to be mechanically picked and had a shelf life of three weeks, leaving plenty of time for shipping and selling. The researchers could not believe their good fortune.

No one looking at the new tomato, which Calgene named Flavr Savr, would have any idea what process had been used to develop it. And since the government had decided not to regulate the process, Calgene was not required to perform premarket tests or even to notify the government of its intention to market the magical new food. Companies were encouraged to consult voluntarily with the FDA. However, Calgene executives looked at the gathering storm over GM foods and concluded that the only way to capture the public's confidence was with full disclosure. There had been a couple of unnerving events in recent months, enough to scare any company thinking of marketing a biotech product.

One occurred at the University of California, Berkeley, where agricultural experts had been looking at ways to prevent frost damage on crops such as strawberries and tomatoes. Most fruits are damaged or destroyed by frost as ice forms on their surfaces. However, ice forms more readily on a regular surface. The bacteria that live naturally on the surface of fruits generally allow ice formation. So the Berkeley researchers wondered if they could find and deactivate the gene in a bacterium responsible for allowing ice to form on delicate fruits and berries. The altered bacterium could then be sprayed on, say, strawberries, produce a rough instead of a smooth surface, and delay frost damage. In short order a product appeared named Frostban, marketed as the "ice minus" antifreeze bacterial spray.[17]

The Environmental Protection Agency allowed a field test in 1987 that attracted a lot of media attention. Antibiotech activists protesting the genetic engineering of the bacterium pulled up most of the strawberry plants the night before the test. The next day a scientist, dressed somewhat unfortunately in a moon suit, sprayed the bacterium onto a small patch of strawberries that had survived. Frostban apparently worked, but the company dropped the project after deciding that opposition to GM products was already too widespread. They would never be able to market Frostban without a fight.

A second exotic defrosting idea was to insert the antifreeze gene from the Arctic flounder into a tomato, endowing it with frost damage resistance. This idea fit perfectly into the "Frankenfoods" category, and opponents of the technology lost no time in publicizing the horror of a "smelly" fish gene in a tomato, as though it might now have gills and a tail. Although the idea never left the laboratory, it took on a media life of its own and became an antibiotech icon.

Another case that raised questions but never provided enough answers involved a widespread health food supplement from Japan named L-tryptophan. Supposed to relieve anything from stress to premenstrual pains, a batch of the drug resulted in the deaths of two dozen people and left fifteen hundred seriously ill. It was not clear how the toxins had appeared in the drug, however. The company had used genetically engineered bacteria in the production of the drug; opponents of the technology quickly suggested that here was a definitive case of some unexpected, unpredictable consequence of the GM process. But some of the same toxic compounds were found in batches of unmodified bacteria used by other manufacturers of the product. It was suggested that the problem might be a new filtration system, recently installed in the production line by the Japanese, that had somehow let through toxic impurities that had nothing to do with genetic engineering. The exact cause of the illness was never discovered. The product was withdrawn from the market, and the Japanese company destroyed the batches of the modified bacteria, making impossible any independent analysis.

As eager as Calgene was to bring the first transgenic food to market, the company was also mindful of such warnings. Former Calgene researcher Belinda Martineau would write in her book *First Fruit* about the development of the Flavr Savr tomato, "Calgene management worried that the questions lingered among other Jurassic Park–inspired doubts in the collective public conscience [and decided that] proceeding conservatively yet quickly to gain approval from the FDA was perceived to be the best approach." [18]

Like Monsanto, Calgene sought some kind of governmental seal of approval to ease the tomato's move into the marketplace—if possible a separate declaration from the U.S. Food and Drug Administration that their tomato was safe to eat. The company was not worried that their new tomato was unsafe, since they were not inserting a new gene or making any new protein. Quite the reverse, in fact. They had stopped protein from being made by reversing a gene. But both Calgene and the FDA were on trial as far as the public was concerned.

Calgene's tomato was the first transgenic whole food ever presented for inspection.

Despite their lax rules, the FDA was also cautious. The agency was concerned about three things. First, they wanted to make sure that the new tomato would be as nutritious as the old one; that the new process would not somehow leach the vitamins from this favorite American food, normally rich in vitamins A and C. The FDA was satisfied with figures provided by the company that the vitamins remained intact.[19]

Second, the FDA demanded a specific assurance that the genetic engineering process had not caused an increase in the level of toxins through some unexpected effect of inserting the reversed gene. Publicly the company had adopted the corporate mantra that genetic engineering was more "precise" than traditional breeding, and therefore likely to produce a safer product. But the researchers at Calgene knew about the limitations of the theory of precision in genetic engineering. They were concerned that they had "no control over where our genes went into the plant's DNA. They could be integrated harmlessly or smack dab in the middle of an existing gene, disrupting that gene's function."[20]

The only naturally occurring toxin in tomatoes is tomatine, one of the glycoalkaloids, a group that includes solanine, the agent responsible for the poisonous green patches on sprouting potatoes. If potatoes have not been fully covered by earth before being harvested, they often have telltale green patches of solanine.[21] Solanine poses a special risk because it cannot be deactivated by cooking, as can toxins in dried beans, for example. There have been cases of solanine poisoning.

Tomatine is estimated to be about one hundred times less toxic than solanine, and Calgene could find no cases of tomatine poisoning. However, studies showed that tomatine levels decreased dramatically almost to zero as the tomato ripened. Calgene researchers were concerned that by tinkering with the PG gene, they might have somehow inactivated a gene responsible for eliminating tomatine during ripening. The FDA demanded that the tomatoes be tested on rats for what is called an acute toxicity test. In such tests, laboratory rats are fed large doses—in this case, the equivalent of a two-hundred-pound person eating three pounds of fresh tomatoes at one sitting. The Calgene tomato passed the second of the FDA's tests.[22]

Finally, the FDA wanted to know more about the marker gene in Calgene's tomato—the one aspect of the new tomato that Calgene was most worried about. All biotech companies were using a marker gene, yet no one had ever considered asking for government approval for this essential part of the process of genetic engineering.

Inserting the gene cassette into the host cell is such a hit-and-miss affair, with such low rates of success, that researchers need some way of knowing that the cassette has arrived successfully. Marker genes provide such identification. By 1991, the company had carried out, as Martineau would put it, more than twenty-one thousand individual transformation "experiments"—trying to transfer the PG antisense gene into several tomato varieties. Only one hundred and thirty transgenic tomatoes produced enough seed for the company scientists to work on. Only forty-four plants were selected as worthy of breeding.

In early experiments researchers used a marker gene from fireflies that produced the enzyme luciferase, which makes the firefly's tail glow. Plants that had successfully incorporated the gene cassette could be distinguished. Later, researchers started using a gene that confers resistance to a commonly used antibiotic. One of the most popular antibiotic resistance genes, known as nptII, produces an enzyme that inactivates the common antibiotic kanamycin. The test is disarmingly simple. The experimental plant cells are grown in a medium containing the antibiotic. If the plant cells have received the antibiotic-resistance gene, they survive. If not, they die.

The potential problem comes later. Once the transgenic plant has grown, the antibiotic gene no longer performs any useful function, becoming excess baggage, but the plant continues to produce the antibiotic-resistance enzyme. Anyone who eats the plant will also eat the enzyme. In theory, the enzyme could deactivate the antibiotic function of kanamycin in human beings who eat the food, thus reducing the drug's therapeutic value.[23]

Scientists were curious whether foods containing the marker gene could allow transfer of those genes to naturally occurring, harmless bacteria in the human intestine. The resistant genes in the harmless bacteria could, in turn, be transferred to bacteria that are harmful. The result could be a harmful strain of bacteria that was resistant to an important antibiotic or even to a whole family of antibiotics.[24]

The companies argued that the chances of the genes being transferred to these bacteria were so small that there was no need to be concerned. They also argued that the risk of antibiotic resistance spreading to bacteria in the human gut was much greater from the heavy use of antibiotics to prevent diseases in cattle and poultry, as well as overprescription by family doctors.

In its submission Calgene noted that large numbers of bacteria isolated from humans were already resistant to these antibiotics. "A fresh salad with lettuce, carrots, celery, cucumbers, and tomatoes is actually a major source of these organisms."[25] Drinking water was another significant source of resistant bugs.

However, the nptII gene presented Calgene with an especially tricky public relations problem. The gene had been isolated from a particular strain of *Escherichia coli,* commonly known as *E. coli,* a bacterium that is widely used in genetic research and occurs naturally in the intestinal tract of humans and animals, as well as in soil and water. Some strains are harmful, causing diarrhea or more serious gastrointestinal infections. Despite all the precautions taken to clean and pasteurize foods, several thousand cases of *E. coli* poisoning occur each year in developed countries, and there are occasional outbreaks in which people die.

Calgene's nptII was also part of what is known as a *jumping gene*—a mobile segment of DNA that moves around within the genome, either by physically inserting itself at various different sites or by producing a copy of itself. "The bottom line was that our [nptII gene] not only conferred resistance to antibiotics used therapeutically on humans but also had been part of a jumping gene isolated from a bacterial species the public knew and feared," Martineau would write. "Any PR related to these particular facts could not be good."[26] Calgene did not mention in public that the same marker gene was in every prototype transgenic plant it was working on at the time, including tomato, cotton, and canola. It was also in most of the other experimental plants of other companies. "We did not discuss these issues aloud. Rather, we seemed to silently agree that there was no looking back," Martineau later recalled.

According to FDA procedures, the company had two options: consider the nptII gene as an additive and submit it to mandatory approval by the FDA, or simply seek an "advisory opinion," which was all the FDA actually required. Biotech opponents were insisting that the marker genes should be regulated as additives. Most of the researchers at Calgene agreed that the marker gene was indeed an additive. The gene made an enzyme that had not been there before. If not the gene itself, then at least the enzyme it produced in the new tomato should be considered as an additive. But Calgene's regulatory staff balked; food additive petitions took three years to get approved, and Calgene was in a hurry.[27] It asked for an advisory opinion, a more general request in which the marker gene could be referred to as a "processing aid." This was normally a faster route to approval, and none of the safety data had to be published—or included on food labels.

Looking at the possibility that the nptII gene could become harmful once it reached the human gut, the FDA wanted to know how many nptII genes would be in the Calgene tomato when it arrived in food stores. As it turned out, that was a problem. No one had actually considered counting the number of nptII genes. Other companies using the marker gene had hoped no one would ask—in case the FDA decided to impose limits and the limit could not be met, making approval impossible. Most of the new transgenic foods—tomato soup, canola oil, margarine—would be intensively processed before being eaten, and genes, including the nptII gene, would be destroyed by crushing and high temperatures. But what about fresh tomatoes?

Calgene estimated, as best it could. For every one thousand people who ate the company's tomato, one bacterium that normally resides in the human digestive tract might become resistant, it concluded. And out of a population of one trillion bacteria normally in the human system, "one didn't seem too bad." [28] In the end, seeing no reason to limit the number of kanamycin-resistant genes or the amount of protein they produced, the FDA was satisfied.

On May 18, 1994, after three years of submissions, corrections, additions, and an open hearing about Calgene's tomato, Kessler's FDA gave the company the go-ahead. But the approval said only that the new tomato was "substantially equivalent" to its predecessors. There was no "safe" label specifically for the Flavr Savr, as Calgene had wanted.

After the long review, Rebecca Goldburg, senior scientist with Environmental Defense, a New York–based green group, admitted that the FDA had "done a considerable review." But Goldburg was still concerned that the FDA was asking for a voluntary system, leaving it up to the companies to decide whether a new product needed review. There was no guarantee that companies developing future products would be equally responsible. Goldburg was not looking far into the future, she was staring at products waiting in the wings. Monsanto had a long list—cotton, potato, and corn, all with a pest-resistant gene, and soybean and canola that would be resistant to Monsanto's own new powerful Roundup herbicide.

The doctrine of substantial equivalence now became part of the arsenal of the antibiotech forces. To Rifkin, the day the FDA gave approval to Calgene was the day he declared his own "tomato war." [29] He vowed to picket markets, hand out notices, and organize tomato dumpings and boycotts. [30] Calgene executives were defiant. "Now they're trying to scare consumers," complained the company's boss,

Roger Salquist.[31] That was exactly what they were doing, and there would be a vast assortment of scary stories to come.

As for Calgene's tomato, the wizards at research turned out to be hopeless at marketing. The tomato varieties they used tasted better, but not that much better. Monsanto watched as the company failed to make a niche in the market, and then it pounced. In the summer of 1995 Monsanto bought Calgene, not for its tomato but for its other bright ideas about genetically modified cotton and canola. The Flavr Savr tomato disappeared.

The arrival of golden rice in the spring of 1999 brought a fresh wave of protests about antibiotic marker genes. The patent search revealed that Potrykus and Beyer had used the nptII gene as a marker for their rice transformations; its use was complicated by a patent owned by Japan Tobacco. A second antibiotic-resistance gene also used in golden rice was covered by a patent owned by Eli Lilly. By then, however, the Europeans had already concluded that genetic engineering could and should proceed without such unpredictable processing aids. The possibility of creating new strains of antibiotic-resistant bacteria might be tiny, but why risk it when there were other marker genes available?

In 1998 a British government report suggested that the use of these genes be phased out, and a year later the British Medical Association recommended that the antibiotic gene should be banned outright. By 2000 the Swiss agbiotech company Novartis had found a substitute that allowed researchers to learn, by adding a simple type of sugar to the growth medium, whether plant cells had absorbed a new gene. While the Swiss company did not concede that antibiotic markers were a real risk, its executives were reacting to the antibiotech forces. "If customers are frightened of something, it is risky for a producer to produce it, and it is riskier for us to sell it," the company said.[32]

By then U.S. scientists and doctors had also started to change their views about marker genes, agreeing that there were better ones available that were not antibiotics. One of these was a jellyfish gene that produces a fluorescent green glow under ultraviolet light.[33] But the FDA stubbornly held back, issuing a cautionary guidance on the genes while still allowing the companies to use them.

THE BATTLE OF BASMATI

Every aspect of the innovation embodied in our indigenous
food and medicinal systems is now being pirated and patented.
—VANDANA SHIVA, BBC REITH LECTURE, 2000

No one is certain where man, or more likely woman, began shoving
young green shoots into the shallow marshes or upland meadows of
Asia in order to cultivate a better supply of rice. The innovators may
have been the nomads of the Greater Punjab, in the foothills of the Himalayas and the tributary valleys of the Indus that straddle northern
India and Pakistan. Or possibly the first rice farmers worked the upper
reaches of the Irrawaddy in Myanmar; perhaps they even labored in a
sheltered basin somewhere in Thailand.

Nor is anyone certain how the two main types of cultivated rice
evolved, whether the temperate strain *japonica* came from the tropical
strain *indica,* or vice versa, or whether they evolved independently
from their common ancestor in the wild, *Oryza sativa*. In contrast, rice
experts speak with absolute certainty—and great reverence—about
the origins of the two exotic varieties of aromatic rice: basmati, which
was first cultivated in the foothills of the Himalayas, and jasmine rice,
which comes from Thailand.

With the arrival of plant patents, the pedigree of these scented
grains became more than an academic curiosity. Suddenly their beginnings, their historic roots, became the focus of a bitter global fight. At

issue was whether the names basmati and jasmine belonged exclusively to rice grown in the Himalayan foothills, or in Thailand, and whether these varieties were distinct enough to be worthy of an international appellation such as Scotch whisky or French champagne. Rice farmers in India, Pakistan, and Thailand, who had never doubted that they were the only ones who could produce these savory plants, found themselves defending not only the name but also the genetic makeup of the rice they had been cultivating for centuries.

To the consumer of rice in Asia, basmati and jasmine are so distinct from each other, and from their common rice cousins, that in earlier times they were considered to be gifts from the gods. Indeed, even the American rice eater, given a taste of real basmati instead of the lifeless white mush produced by the prepackaged grains at the grocery store, would agree that it is no ordinary mortal dish. Compared with common rice, the basmati grain is slender, fragrant, and translucent, often with an opaque white halo at its tip. The word *basmati* is derived from the Sanskrit—*vas,* meaning "aroma," and *mayup,* meaning "ingrained" or "present from the beginning." The earliest mention of the seductive basmati was in the epic *Heer and Ranjha,* composed by the Punjabi poet Varish Shah in 1766.[1]

Some of the best basmati once came from the Indian village of Tapovan at the top of a hilly ridge near Rishikesh. The rice was planted in an ancient meadow surrounded by sacred mountains. The nutty aroma of Tapovan basmati was so strong and distinctive that people in neighboring villages always knew when the Tapovanis were preparing their food. At one time the local monarch, the King of Tehri, decreed that only basmati should be grown in Tapovan and that the entire crop should be given to his royal family. When ownership of the Tapovan land shifted to the head priest or *mahant* of Bahat Mandir in nearby Rishikesh, the entire crop of basmati from the village was given as an offering to the temple, as too good for anyone but the local deities.[2]

By the nineteenth century, Indian nobility would eat only basmati rice; a sack of basmati was even considered a suitable present for diplomats to carry to foreign courts. The rice was especially favored by the Persians, even though they grew their own aromatic types. Basmati was an essential ingredient in the exquisite and elaborate Moghul casserole, known as biryani, alternate layers of saffron-flavored rice and lamb cooked with several different spices, the entire dish garnished with crisp sautéed nuts, crackling onion shreds, a splash of rose water, and edible pure silver foil. One famous biryani recipe was named after the Moghul emperor, Shah Jahan, who built the Taj Mahal.

With such a pedigree, basmati was always in demand. Rice farmers of the Greater Punjab vied with each other to breed the finest varieties. By the early 1900s basmati fetched double the price of ordinary rice and became a prize agricultural export worth hundreds of millions of dollars annually. The rice became known as the "scented pearl" of Asia.

It is not surprising, then, that Indian and Pakistani farmers were extremely upset in the late 1990s when a small Texas rice company, RiceTec, Inc., claimed to have bred a new variety of basmati "similar or superior to" Asian basmati. Worse, the company was awarded U.S. Patent no. 5,663,484,[3] which made twenty claims on three new basmati breeding lines developed by crossing dwarf Green Revolution varieties with basmati grown traditionally in Pakistan. RiceTec planned to market "traditional basmati style" under the name Kasmati, and "American basmati" under the brand Texmati.

"Biopiracy," cried Vandana Shiva, the Indian environmental activist, using the latest war cry against biotech agriculture.[4] The word was invented to describe companies such as RiceTec that went hunting for exotic genes in undeveloped countries and then borrowed those genes to breed their own patented versions of local crops. The companies themselves preferred to call their activities "bioprospecting."

In its defense, RiceTec claimed to be a small operation whose production of American basmati would represent a mere blip in the worldwide basmati market. Moreover, the patent was only valid in the United States, the company argued. RiceTec's monopoly control was confined to the production of basmati in North America.

India, however, saw nothing less than a Texas firm trying to rustle away another bit of Indian culture. The government in New Delhi decided to mount a challenge to the patent—a decision never taken lightly by developing countries because of the huge cost involved. In recent years Indians had grown increasingly disturbed by the way foreigners from pharmaceutical interests had been patenting their traditional plant medicines and adapting their local herbs for international profit. As far as many in India were concerned, basmati was another example of this high-tech biopiracy.

Americans had filed a patent on the medicinal properties of turmeric, the bright ochre spice that had been a basic ingredient of Indian cooking and healing since the beginnings of Indian culture. In 1995 University of Mississippi medical researchers received a U.S. patent on turmeric. The yellow powder is obtained from the roots of *Curcuma longa,* a herbaceous plant native to India and Southeast Asia. The U.S. Patent Office granted a patent to the Mississippi researchers for "a method of promoting healing of a wound by administering turmeric to a patient afflicted with the wound." U.S. patent law recognizes "prior art" in foreign countries as one factor that disqualifies a patent claim; India presented an ancient Sanskrit text as proof of prior art. Three years after the turmeric patent was issued, it was revoked.

Another U.S. company—this time a big multinational, W. R. Grace—had patented the antifungal agents in the seeds and the bark of the neem tree, which grows all over India. For centuries Indians have been using the bark of the neem tree, whose name means "free tree," to kill insects and pests, but also as a disinfectant. Indians eat the leaves to build up antibodies. They massage their gums with neem twigs or buy a native toothpaste sold under a neem brand name. Neem, the get-well potion of India, has been touted as a cure-all for such ills as leprosy, snake bites, smallpox, insomnia in babies, and hysteria.

At the same time, European companies were also developing their own versions of neem tree medicines. During the 1990s more than ninety patent applications were received by the European Patent Office for "inventions" based on the neem tree. Eleven were granted, including the one from W. R. Grace. Vandana Shiva led a challenge against those patents that was successful in Europe.

All the patents had been drawn up by a growing army of clever patent lawyers, mostly American, who were filing increasingly complex and broad claims. By the time the patent office was hearing details of the basmati case, RiceTec and its lawyers had figured out how to put the Indian basmati connoisseurs on the spot.

There are many types of aromatic rices, but basmati has it own special scent. A basmati farmer or trader can easily distinguish a real basmati aroma from its lesser competitors. In fact some experts believe that the scent is so strong that the normal human nose can detect the chemical compounds that create the smell even diluted to one part per billion.[5] But the exact makeup and the mix of the chemicals and the physical conditions needed to produce the basmati aroma are still a mystery. Temperature is a factor. The most pungent of the basmati rices grow in fields where the temperature is relatively cool during the early growing season. Basmati grains are also thinner and a few millimeters longer than ordinary rice grains, and the length is also affected by temperature. In the warmer climates of Pakistan, the basmati grain is shorter than the basmati of the Indian Punjab, where the growing season is cooler.[6]

Basmati rice has long been popular in Europe, but in the United States it is a newly acquired taste. Farmers in the South started to experiment by growing some basmati, but the varieties were not popular.

When RiceTec came along, the American domestic market was open. In its patent claims, RiceTec distinguished its varieties by scent, of course, but also by what was referred to as a "starch index." Rice grains are almost all starch, but there are two main types, amylose and amylopectin. The ratio between the two determines whether the cooked rice is chalky, with grains that grow longer with cooking, or sticky, with grains that remain the same length in cooking. Basmati is high in amylose and thus chalky with longer grains once it is cooked.[7]

RiceTec claimed a higher amylose ratio in its varieties of basmati—and argued that therefore they had created a higher-quality basmati. India viewed the evidence of the starch index largely as a clever legal trick, a maneuver designed to persuade the patent examiner that the company had discovered something new, when in fact basmati's grains had grown longer as they cooked for Indian families going back several millennia. The Indians believed that concentration on the starch index diverted attention from the real issue, the use of the word *basmati*. India argued that only rice grown in northern India and Pakistan deserved to have this name.[8]

India's challenge was partly successful. In June 2001 the United States Patent Office rejected 15 out of the 20 characteristic claims under RiceTec's patent. In a forty-six-page ruling, the patent office said that the rice lines, plants, and grains that RiceTec had claimed in the patent to be new were in fact "substantially identical" to basmati varieties grown in Indian and Pakistan. The company had the right to appeal.

Hard on the heels of the RiceTec battle came the skirmish over basmati's aromatic peer, the highly prized jasmine rice from Thailand. In the fall of 2001, Thai farmers heard that two American researchers from the University of Florida's Institute of Food and Agricultural Sciences had developed a strain of jasmine rice that they claimed was suitable for growing in the United States. Thousands of Thai farmers, protesting outside the U.S. embassy in Bangkok, burned President George W. Bush in effigy. They were worried that the Florida researchers' new strain of jasmine would be patented and grown throughout the United States, threatening Thai exports. About five million Thai families grew jasmine rice, which accounted for 25 percent of Thailand's total rice export, 90 percent of which went to America. This was big business for Thailand. Regular varieties of American-grown rice fetched about $340 a ton, but jasmine rice from Thailand sold for $520 a ton, a difference of 44 percent. In the end, the Florida researchers did not develop a strain of jasmine rice, but the Thais fear that one day somebody will do so.

The highly publicized legal tussles over turmeric, the neem tree, and basmati and jasmine rice fueled a bitter debate over the way patent laws allow companies to assume ownership of the knowledge of indigenous people from developing nations. A UN report called this gene drain the "silent theft of centuries of knowledge."[9] The knowledge being lifted by international seed conglomerates included patents on tea, chutneys, coffee, pepper, cauliflower, cabbage, peas, melons, and hallucinogenic vines. Still, the real worry was that some conglomerate would find a way to recreate and then replace agricultural products once found only in tropical species. A British food company had two patents on the flavor gene from a West African cacao tree, which in theory could be used to produce cocoa artificially. The next step would be to produce a substitute for chocolate that bypassed the need to purchase cocoa beans from Latin America or Africa. And then another multinational food company might patent caffeine genes and produce a substitute for coffee, putting Kenyan coffee growers out of work. In this version of modernization through genetic wizardry, the losers would be the poor farmers whose wisdom came from earlier generations, not from a new class of molecular biologists.[10]

The international debate over biopiracy, or bioprospecting, pitted corporations against developing countries' governments and indigenous peoples. The thrust of this diplomacy was to find ways to eradicate the perceived social injustice of rich northern countries' taking natural resources from a poor country in the South and making a profit. One UN estimate suggested that plants and animals taken from tropical countries were worth more than $20 billion a year to major pharmaceutical companies.[11] A report commissioned by Christian Aid estimated that biopiracy was costing Third World countries $4.5 billion a year.[12] Participants in the debate hoped to find some middle ground, a way of compensating the peasant farmers whose ancestors had nurtured the exotic plants over centuries as well as rewarding the biotech researchers for their costly efforts at turning those plants into artifacts that benefited humankind. But the issues were not easy. One question was whether a country could own the natural resources found in tropical forests, or whether they should be considered the "common heritage of mankind"—the concept that arose in the Law of the Sea debate.[13] Another question was the extent of the global responsibility of corporations. In the view of Nigel Dower, a sociologist at the University of Aberdeen in Scotland, it was "crucial to recognize that, given the evils of extreme poverty and lack of development, there is a global responsibility [of corporations] to facilitate and not impede development." If bioprospecting and the patenting that inevitably followed hurt development in a Third World nation, then, said Dower, "we all have a responsibility to modify patterns of activity."[14] Yet another question was whether there was a difference between the old, familiar trade in raw materials, such as tin, bauxite, and oil, and the new trade in living things.

Even the companies agreed that there was an urgent need to overhaul the current international patent system to end obvious abuses and to prevent the issuing of broad and overlapping patents such as had derailed the independence of the golden rice experiment. The system had fallen into disrepute after the landmark 1980 U.S. Supreme Court decision that living cells were patentable. Before that time there had been no patents on plants or plant breeding, only a relatively weak form of legal protection known as "plant breeders' rights." Breeders still happily swapped the product of their labors for research, and farmers were proud to share with their neighbors the seeds of a new prize corn or wheat. But the Court's decision had turned the familiar world of plant breeding upside down.

When the first nomads reached North America, the continent had a tiny range of edible plants, at least for humans. A small assortment of berries, a few tasty sunflower seeds, and the Jerusalem artichoke made up much of a sparse fare. European settlers, of course, brought their favorite foods and, in order to encourage Americans to keep adding new species of food plants to the national store, Thomas Jefferson famously declared that the "greatest service which can be rendered to any country is to add a useful plant to its culture." Plants were an obsession of Jefferson's, but he also wanted Americans to bring back new gadgets and machines from Europe. With the U.S. Patent Act of 1793, which was one of Jefferson's many achievements, he offered protection for "any new and useful art, machine, manufacture, or composition of matter, or any new or useful improvement thereof."

From the beginning, a U.S. patent was granted if the inventor could show that the invention had never been made before, involved a "nonobvious" step, and served some useful purpose. Certain discoveries could not be patented: physical phenomena, abstract ideas, and the laws of nature. A new mineral discovered in the earth or a new plant found in the wild could not be owned, even temporarily. Such discoveries were "manifestations of nature, free to all men and reserved exclusively to none," the act said. The act embodied Jefferson's Lockean philosophy that "ingenuity should receive a liberal encouragement." 15

In the early nineteenth century the first wave of professional plant hunters set off in search of suitable germ plasm to fill America's larder. Seeds were brought back from distant lands, bred on state-run agricultural research stations, and distributed free to farmers. When it became clear that there might be profits in the seed business, attitudes toward protection of privately developed varieties changed. The commercial seed companies demanded protection for their wares. By the end of the nineteenth century the role of the public agricultural agencies was seen increasingly as "an institutional impediment to the expansion of private enterprise in the seed business."[16]

Breeders began calling for the establishment of a plant patent system, claiming that it would encourage the breeding of superior varieties. Their rallying cry was, "Every seed is a mechanism as surely as is a trolley car."[17] But it would take another generation before breeders received any rights over their prize varieties—and then only a relatively weak form of legal protection. America's most celebrated plant breeder, Luther Burbank, complained to the House of Representatives: "A man can patent a mousetrap or copyright a nasty song, but if he gives to the world a new fruit that will add millions to the value of the Earth's harvest, he will be fortunate if he is rewarded by so much as having his name associated with the result."[18]

In 1930, Congress passed the Plant Patent Act, but it covered only asexually propagated species, such as fruits, nuts, and flowers. Congress was reluctant to give breeders monopoly control over staple food crops. Potatoes, for example, were specifically excluded. U.S. breeders continued to push for wider property right protection—especially after European nations created an international plant patent–like protection system in 1961.[19] The new system, known as the International Convention for the Protection of New Varieties of Plants, was a compromise recognizing the breeders' need for protection and the fear that actual patents on plants could lead to monopolies and drive up the price of food and seeds. These plant breeders' rights permitted other breeders to use the protected varieties as source material for their breeding programs, and, equally important, farmers were allowed to keep the seeds for replanting next season. The new plant breeders' rights did not solve the problem of whether biological products could be patented—a topic of increasing debate in pharmaceutical and plant breeding circles.

In the United States, a similar law—the Plant Variety Protection Law—was passed in 1970, giving patentlike protection to food crops that reproduced sexually—the staples such as corn, wheat, and rice. Much as breeders complained about their inferior property rights status, this was really all the protection they could expect. Until the advent of biotechnology and DNA analysis, verifying the parentage claims of new distinct varieties was problematic. Claims over heritable traits would have been unenforceable. In any case, the American breeders had a good deal. They had obtained effective property rights without having to submit their new varieties to government testing for quality. In contrast, European plant breeders had to demonstrate to government inspectors that their new varieties were superior to older varieties before they could obtain a certificate.

Then the U.S. Supreme Court made its fateful decision to expand the class of patentable subject matter to new life forms. In 1972 a General Electric microbiologist, Ananda Chakrabarty, filed a patent application for a bacterium that would soak up oil spills. Chakrabarty had developed the microbe by putting different strains of bacteria together in a laboratory culture—a kind of bacterial soup—that allowed the microbes to exchange genetic material, just as they would in nature. The result was a strain that digested oil, a feature possessed by no known naturally occurring bacterium. At the time it was an especially attractive feature. Oil tankers were frequently running aground and producing massive oil slicks. But Chakrabarty's application was turned down by the patent office on the grounds that living things were not patentable.

General Electric appealed and in 1980, in a five-to-four decision, the Court held that Congress had meant the patent act to have wide scope—anything "under the sun" made by man's ingenuity was patentable, including Chakrabarty's bacterium. While the majority opinion still held that a new mineral or a new plant could not be patented, the fact that the bacterium was alive was "without legal significance" for the purposes of the law. Chakrabarty's new microbe had "markedly different characteristics from any found in nature and one having the potential for significant utility."

The Court's dissenters argued that in the two plant protection laws of 1930 and 1970, Congress had addressed the general problem of patenting animate inventions and had chosen specific language to exclude them.[20] In fact, the dissenters noted, "Congress specifically excluded bacteria from the coverage of the 1970 Act. The Court's attempts to supply explanations for this explicit exclusion ring hollow." In any event, the dissenters argued that it was the "role of Congress, not this Court, to broaden or narrow the reach of the patent laws."

Critics of the Court's decision stressed that Chakrabarty had not actually created anything new, he had merely intervened in the normal way bacteria swap genes. One National Academy scientist, Key Dismukes, observed that Chakrabarty's strain "lived and reproduced itself under the forces that guide all cellular life. . . . The argument that the bacterium is Chakrabarty's handiwork and not nature's wildly exaggerates human power and displays the same hubris and ignorance of biology that have had such a devastating effect on the ecology of our planet."[21]

Chakrabarty himself was surprised by the Court's decision, since he had simply cultured different strains, hoping that they would swap their genes in the natural way. "I simply shuffled genes, changing bacteria that already existed," he said. "It's like teaching your pet cat a few new tricks."[22] What made the decision especially galling to those who thought it was simply wrong, either scientifically or morally, was that the bacteria never worked for oil slicks.

Wall Street loved the Court's ruling. A few months later the stock price of the leading biotech firm, Genentech, increased from thirty-five to eighty-nine dollars in twenty minutes during the company's initial stock offering—before the company had introduced a single product onto the market. Genes would become the "raw resource for future economic activity," observed Jeremy Rifkin.[23]

There was no immediate rush to patent plants, since breeders still had rights under the 1930 and 1970 acts. In 1985, however, the patent office overturned a half-century of federal policy and granted a series of patents on a new line of corn to a Minneapolis microbiologist, Kenneth Hibberd. His biotech company was granted patents on the tissue culture, the seed, and the whole plant. At the time, plant breeders wanting to register their inventions could still use the Plant Variety Protection Act, which covered a single claim for a new plant variety. But the Hibberd patent ruling gave them a distinct advantage. It covered the all-important process of creating the new variety as well as the product—the DNA sequences, genes, cells, tissue cultures, seed, specific plant parts, and the entire plant. (These were the same kind of process patents that had caused Potrykus and Beyer so many problems in creating their golden rice).

The Hibberd ruling also encroached upon the farmer's right to plant seed from his harvest. In the existing plant protection laws, there was a farmer's exclusion clause allowing farmers to plant seeds from the protected variety. In contrast, the purchase of a patented seed gave the farmer the right to grow the seed, but not to save and replant.[24]

Biotech companies started to file very broad patents that, if taken at their word, could bring under their monopoly control the key techniques in the emerging genetic engineering of plants. The U.S. Patent Office began "routinely issuing patents on products of nature (or functional equivalents), including genes, gene fragments and sequences, cell lines, human proteins, and other naturally occurring compounds."[25]

The big change was that patents were issued on products of skill rather than invention. One company was granted a patent on a compound, which it had extracted from human urine, that stimulated the production of red blood cells. Another company received a patent on a blood-clotting agent that it had extracted from human blood. The U.S. biotech industry, strongly supported by the Reagan administration, was on a mission of national importance. They were in a "global race against time to assure our eminence in biotechnology," and the U.S. Patent Office would play an obliging role.[26]

In the same five-year period, Congress significantly expanded the class of possible patentees with the passage of the 1980 Bayh-Dole Act. The new law allowed university researchers to take out patents on federally funded projects.[27] Supporters of the act argued that important inventions would languish in university laboratories without such legislation; critics foresaw the free exchange of scientific material drying up. The universities themselves argued that the patents provided royalties that would keep their research laboratories alive at a time of drastic cuts in public funds. Many university patents and a number of successful products resulted, but academic researchers who used to swap inventions as freely as farmers used to swap seeds now began to take property rights into account.

The message from Congress was clear. University researchers could, and even should, take advantage of available private funds. Researchers began to set up deals with private companies. Initially there were cultural clashes that made researchers cautious. But in 1995 the companies became bolder in their research objectives, pushing to control all the elements of biotech—seeds, genes, and genetic information—and the researchers also became more daring.

The most controversial college deal was struck in 1998, when Berkeley's Department of Plant and Microbial Biology teamed up with the Swiss biotech and agrochemical giant Novartis. The company agreed to pump $5 million a year for five years into the department and give researchers access to their confidential databases in return for first rights to negotiate a license on any new invention the department produced. Academic researchers were striking similar one-on-one deals with corporations, but the Berkeley agreement was unusual in that it involved virtually the entire department. All but two of the nineteen researchers signed on. Critics complained that the department had sold out its independent academic mission to corporate interests. Legislation was now in place in the United States for the rapid demise of publicly funded agricultural research.

The number of biotech patent applications skyrocketed. U.S. patents for rice plants, for example, rose from fewer than a hundred to more than six hundred a year between 1995 and 2000. Most of these patents were privately held. One survey found that about three-quarters of plant patents were held by companies, with nearly half in the hands of fourteen multinationals.[28]

The new patent regime also led to a major restructuring of the seed industry through mergers and acquisitions—often for the purpose of buying up patent portfolios. One company was created solely for the purpose of "buying up broad patents and then suing other companies for alleged infringements."[29] Between 1995 and 1998 Monsanto, DuPont, and Novartis had spent $30 billion acquiring other agbiotech companies.[30] By the end of 1998 Monsanto had been involved in eighteen acquisitions. Novartis was formed by the merger of Sandoz and Ciba-Geigy. DuPont set up joint ventures that were worth more than $5 billion.[31]

At the beginning of the twenty-first century, the concentration of agbiotech patents in the hands of five conglomerates made genetic engineering more difficult for the smaller biotech companies and independent college researchers. A California biotech company, DNAP, had used a promoter from Monsanto in its new slow-ripening tomato called Endless Summer. But Monsanto had just bought Calgene and its Flavr Savr tomato. Monsanto refused DNAP use of the promoter and the Endless Summer was closed down.

Independent research was further complicated by the broad patents issued in the United States and also in Europe. To make their clients' patents cover as much of the transgenic process as possible, lawyers wrote them to include "all transgenic cotton plants," or "all genetically engineered soybeans." Rival companies objected, but the bigger corporations always had a way out: drop the costly patent challenge and buy the company. When the biotech company Agracetus was awarded a patent covering genetically engineered soybeans in 1994, Monsanto was outraged and immediately launched a legal challenge charging that one company would have a monopoly over all transgenic soybeans. Monsanto then bought Agracetus and the patent. Now, in addition to the patent on soybeans, Monsanto also owns a patent in both Europe and the United States on genetically engineered cotton.[32]

In others cases, such as golden rice, several patents overlapped. One of the most notorious involved transgenic pesticide technology using the bacterium *Bacillus thuringiensis,* known as Bt. The bacterium produces a toxin that has been used as a natural pesticide by organic farmers for more than half a century. Several biotech companies began to engineer the gene for the Bt toxin into crop plants, filing patent applications as they completed parts of their research. The patent office obliged. The result was hundreds of overlapping patents, many of which are being challenged in court. One of them covers "any insecticidal gene in any plant."[33] At least four different companies have claimed ownership of corn varieties transformed with the Bt pesticide gene. As one patent expert said, it is "almost impossible for a researcher to find ways through this patent thicket."[34]

The proliferation of patents also spelled trouble for the small peasant farmer in a developing country. The danger was that as the big biotech corporations began to take an interest in the staple crops—especially rice in Asia—the publicly funded research institutions, such as IRRI in the Philippines, would find it increasingly difficult to supply poor farmers free of charge with the latest, and best, improved varieties. The oligopoly of the biotech conglomerates sought protection for their process patents, which originally were granted in the United States or Europe, in less developed countries as well. Use of laboratory procedures and the contents of gene cassettes would become more restricted, and available only to those who could afford them.

As John Barton, a patent expert of Stanford Law School, forecast,

"It may be impossible or at least very expensive or difficult for the public sector to gain access to patented technologies or to use protected varieties for research in developing new applications for the smaller crop or subsistence farmer."[35]

One cannot leave this battleground and its apparent injustices without asking two important questions. The first is why the Indian government had failed to apply for an appellation—a Geographical Indicator, or GI—to protect basmati in the same way Scotch whiskey and French champagne are protected. The answer is a lack of political will. A GI would closely define the area where basmati grows and might reduce, or even exclude, business of the wealthy basmati exporters. The second question is more semantic—whether "biopiracy" can ever be an appropriate term. Almost every society has benefited over the centuries from "biopiracy" in some fashion, as the probiotechs like to point out. Corn comes originally from Mexico, potatoes and tomatoes from Peru, cassava from Brazil, wheat from the Fertile Crescent, and so on. Anyone who grows such foods in a country where those foods did not originate—Americans and corn, Africans and cassava, Indians and wheat, for example—is a "biopirate," so the argument goes. In this view, the Texans, who acquired basmati and jasmine rice were only doing what their forebears had done—enriching the food supply—and what the U.S. patent system now allows—making money in the process. One wonders what Thomas Jefferson would have had to say about all that.

Of Cauliflower, Potatoes, and Snowdrops

It's very unfair to use our fellow citizens as guinea pigs.
—Arpad Pusztai, researcher on GM potatoes, 1998

Our confidence in this technology and our enthusiasm for it has, I think, widely been seen, and understandably so, as condescension or indeed arrogance.
—Robert Shapiro, chairman and CEO of Monsanto, 1999

Most people who buy vegetables have probably never heard of the cauliflower mosaic virus, a promiscuous organism found in all members of the cabbage family—including cauliflower, Brussels sprouts, and broccoli. Experts in this virus reckon they can spot with the naked eye evidence of its presence in the mottled pattern of different shades of green that it causes on the leaves of these vegetables, or in the slight discoloration on the cauliflower's curdle of white florets. The virus is harmless, as far as we know. No illness has been reported from its steady consumption since the cabbage family was first cultivated by ancient Egyptians and then brought to Europe and eventually to the New World.[1]

Throughout the history of cauliflower and cabbage, the virus attracted little attention—until the early 1980s. That was when Roger

Hull, a British microbiologist, discovered the virus's genetic secret. Tucked away in its genome is an especially energetic promoter, the 35S. When the virus infects a plant cell, the 35S starts driving genes at an unusually frenetic pace. Bioengineers wondered if the energy of the 35S could be harnessed. A super promoter might be just the answer for switching on the alien character genes they were inserting into various food plants. As it turned out, the 35S worked brilliantly, kick-starting some of the more reluctant genes. Moreover, government regulators in the United States and the United Kingdom passed the 35S as a harmless addition to the bioengineer's tool kit. As a result, the promoter was present in virtually all the first transgenic food plants to come out of the biotech laboratories. Monsanto, the first company to grasp the potential of the superior power of the cauliflower connection, in 1984 applied for an ambitiously broad patent claiming rights to any gene cassette containing the 35S. The claim was granted a decade later— just as the first transgenic food crops were emerging from the biotech laboratories.

In the publicity given to the debut of such wonder foods as the Flavr Savr tomato, weed-killer-resistant soybeans, and new pest-resistant corn, the 35S was hardly ever mentioned. But in 1999 one of the more vocal antibiotech activists in Britain, a geneticist and biophysicist named Mae-Wan Ho, decided to single out the 35S as an example of what was wrong with the new technology. She warned that use of the 35S in transgenic plants was a "recipe for disaster"—that the 35S was so powerful, it could spur into action all kinds of unwanted genes in the food plants, and genes in humans that could cause cancer.[2]

Ho had begun her thirty-year career in genetic engineering at the University of California at San Diego, and since 1994 her opposition to biotech had been relentless. She described the technology as "crude, unreliable, uncontrollable, and unpredictable." For several years she had been a lecturer at the Open University in England, a college for nonresident students. From that platform she provided the antibiotech forces with a stream of doom-laden press releases and scientific papers. At the same time, she lobbied government ministers to call a halt to the use of biotech agriculture before something dreadful happened.

Her basic argument centered on the instability of DNA once taken out of its original surroundings. In Ho's view the "transgenic organism is, in effect, under constant metabolic stress, which may have many unintended effects on its physiology and biochemistry, including increase in concentrations of toxins and allergens."[3] In the gene cassettes used to make transgenic plants, the 35S was artificially linked to genes it had never met before and then shoved into unfamiliar territory. Ho argued that the 35S had "hot spots" that enabled it to hop around the genome of the plants into which it had been inserted, causing the organism unusual stress. The 35S might switch on genes that were not supposed to be activated. It might land next to a dormant viral gene and accidentally wake up a sleeping virus, or it might switch on a gene that could produce some new monster toxin. Finally, she suggested, if the promoter were consumed by animals and humans, it could mix with such human viruses as HIV or hepatitis B with disastrous consequences. Even worse, she warned, the promoter could activate an oncogene, a gene that produces cancer cells, and stimulate that gene to cause cancer.

Ho had bitter encounters with her peers. Some scientists accepted the idea that the 35S might, in theory, move around inside the genome of its host, although not to the extent of other jumping genes better known for such propensities. Other scientists agreed that there might indeed be sleeping viruses in plants, but they dismissed the possibility that such viruses could be awakened by the 35S. The most controversial issue was Ho's suggestion that people might ingest the 35S by eating the transgenic foods and that once inside the human body, the 35S could slip into overdrive and wreak havoc.

Ho's critics, including the 35S discoverer, Dr. Hull, pointed out that humans have been consuming the 35S promoter (from infected members of the cabbage family) at levels more than ten thousand times greater than would be found in any transgenic plant containing the promoter.[4] If consumption of DNA containing the 35S were that dangerous, then humans would have more to fear from vegetables infected with the cauliflower mosaic virus than from transgenic plants. Ho insisted that the numbers didn't matter. When a person eats a cauliflower infected with the virus containing the 35S promoter, she argued, the virus is wrapped in a protective protein coat that renders it not infectious to humans, so eating it "is of little consequence."[5] She argued that when the promoter is used in transgenic plants, it has no such protective coat. It is "naked" and may be more troublesome. But Ho's critics in the scientific community disagreed. They insisted that the virus humans consume from cauliflowers contained both naked and protected particles.[6]

Ho's extrapolations stirred up her peers, but none so violently as the virologists. One expert accused Ho of creating "pure fiction, and lies."[7] When Ho suggested that all transgenic crops should be withdrawn from the fields, and from human consumption, while governments invoked the precautionary principle and did more research, another virologist suggested sarcastically that Ho should have called for a complete ban on all food. "Let's stop eating plants and animals altogether," said the virologist. "It's a shame we did not have [Ho's] information millions of years ago. It would have been so easy to avoid the perils of life."[8]

Other scientists emphasized that Ho was only posing hypothetical questions, that she had not done original research. Two French scientists accused her of mixing science and politics. "Considering the complexity of the debate concerning GMOs, which is not only scientific (risk assessment), but also touches on important questions of an economic, sociological, or political nature, it is essential to take care to make the distinction between scientific questions raised for reasons that are primarily political, and the truly scientific ones that merit devoting considerable effort of analysis and reflection."[9]

Scientists from Britain's John Innes Centre, one of the country's leading biotech laboratories that receives government and industry funding, examined Ho's arguments and concluded that there was no evidence that the virus promoter would "increase the risk over those already existing from the breeding and cultivation of normal crops." [10] In other words, consumers would be no worse off eating golden rice (which also contained the 35S) than they would be eating cauliflower and cabbages from the market. The British researchers reprimanded Ho for not taking the science of biotechnology seriously. "The transgenic situation has to be compared with the natural situation, not with a utopian one." In time the British Royal Society, the nation's senior common room of science, also concluded that there was "no evidence" that the cauliflower virus had caused disease in humans, and that "the risks to human health associated with the use of specific viral DNA sequences in GM plants are negligible." [11]

The report did little to quiet Ho and her colleagues, who continued to rail against the 35S. When golden rice was announced, they attacked the invention as a "useless application, a drain on public finance, and a threat to health and diversity." The new rice "possessed all the usual defects of first generation transgenic plants plus multiple copies of the cauliflower mosaic virus promoter, which we have strongly recommended withdrawing from use on the basis of scientific evidence indicating this promoter to be especially unsafe." [12]

In the consumer's struggle to understand scientific complexities, Ho is an unhelpful tutor, leaving the consumer caught in the middle, mistrusting both her voice and the voices of her critics. The forceful promoter of the cauliflower mosaic virus had been approved by government regulators, and is still being used in transgenic foods. For all the fuss, the consumer is no worse off, apparently—except perhaps for being confused by the debate. Consumers lurched from complete ignorance about such matters as the 35S promoter or antibiotic-resistance genes, to a full-blown panic that the tools of biotechnology might be poisoning them today and tomorrow destroying the means of producing enough food to keep the world's population alive.

While Americans, whether they were aware of it or not, were eating an increasing number of foods containing modified corn, soy, and canola, in Europe a powerful alliance of professional environmentalists, including Greenpeace, Friends of the Earth, and the organic growers of the British Soil Association, were leading a revolt against GM foods. They quickly succeeded in halting the onward march of U.S. biotech crops and blocking the biotech giant Monsanto from its plan to market the new foods in Europe. The battle would be called one of the most "surprising and telling cultural struggles of the late twentieth century."[13]

A mere two years had passed from the day in 1996 when the freighter *Ideal Progress* docked at Hamburg with the first shipment of U.S. genetically modified soybeans for use in a variety of groceries, including margarine, cake, and chocolate, until 1998, when the European Union closed its doors to GM crops. No more GM crops would be authorized for local planting, and imports of GM food grains would be curtailed. The GM controversy devastated sales of U.S. farm products, halving soybean exports to Europe.

The arrival of the *Ideal Progress* launched two years of demonstrations. European protesters began blocking shipments of GM crops in the Netherlands, dumping soybeans in Brussels and Paris, and ripping up test plots in the United Kingdom. Monsanto, the U.S. agribusiness giant that had engineered the herbicide-resistant soybeans, thought the protests would die down. "There is a lot of noise out there by groups trying to make people believe there is something wrong," said a Monsanto spokesman.[14] No reassurances from the scientific establishment about the potential value of the new technology and its inherent safety made any difference. Public confidence in scientists had been shaken by recent food scares, especially "mad cow" disease.

The antibiotech forces divided into three groups. The rejectionists believed that for religious, environmental, or food safety reasons plant biotechnology was wrong, dangerous, and should be stopped. They feared irreversible harm to natural landscapes and biodiversity as well as unknown allergenic and carcinogenic effects. They were dismayed at the widening gaps in wealth and power between North and South and feared the new technology would only expand those gaps.

The second group, the reformists, believed that scientists, businesses, and government had mishandled the new technology; the technology itself was not the root of the problem. They focused on consumer choice through labeling of the new products, more information from the biotech companies, more control over the corporations. They disliked having the new foods "shoved down our throats" and advocated time to debate the issue.

The third group—the organic growers and consumers—demanded immediate labeling, intrusive tracing of the new seeds, and isolation of GM crops to reduce the risk of cross-pollination and genetic pollution.[15]

A common complaint in each of the three groups was the Americanization of European eating habits, as well as its agriculture. Fast-food chains such as McDonald's had long been under attack by radical groups as symbols of unwanted American culture, but the simple fact is that Europeans and Americans have very different eating styles. In Europe, national recipes are closely held—even in Britain, which earned a reputation for greasy, unpalatable fare in the post–World War II years. Europeans spend a much larger proportion of the family budget on food than Americans, in part because food is more expensive, of course, but also out of choice. They tend to take more care over what they buy than hurried Americans sprinting through the grocery store. They spend more time preparing their meals and more time eating them. The Jimmy Buffet hit "I Wish Lunch Would Last Forever" presents so revolutionary a thought in the United States that it is practically un-American. By contrast, long lunches perform an important social function in European society.

In the antibiotech vanguard were the ecowarriors of Greenpeace, whose leaders rose out of the German left in the 1980s—people like Benedikt Härlin. He had been prominent in the occupation of the "liberated zone" of tenement houses near the Berlin Wall and a member of a collective that published *radikal,* a "newspaper for uncontrolled movements." When President Ronald Reagan visited West Berlin in 1982 during the cruise missile crisis, the newspaper helped organize protests against U.S. nuclear forces. Police raided the newspaper and took away files. Its organizers, Härlin among them, were jailed for thirty months for aiding and inciting terrorist acts. Once out of jail, Härlin, then twenty-seven and less radical, won a seat in the European Parliament as a representative of Germany's new Green Party. On a trip to America, he met Jeremy Rifkin, learned about biotechnology from one of its foremost opponents, joined Greenpeace, and helped organize the welcoming party for the first boat into Hamburg with GM soybeans.[16]

From that day, the greens focused their battle on the uncertainties of biotechnology. If they could convince consumers that GM foods were unsafe—for humans as well as the environment—they might persuade European governments to adopt the precautionary principle that required more research. They might even stop the technology. After a series of food scares, European consumers were vulnerable; they began to listen to voices previously regarded as fringe. The most important staging for antibiotech arguments was Britain. In the home of the Luddites, of Blake, Rossetti, and Ruskin, where country lovers in Wellington boots go batty about the disappearance of the skylark, where there is no shortage of organic farmers practicing muck and magic agriculture, the green movement found eager allies.

In 1996, when the first transgenic crops were being harvested in America, Britons had no qualms about buying the first GM product to appear on European supermarket shelves—a tomato paste that was cheaper than its conventional equivalent. At the beginning of 1998, GM tomato paste in U.K. supermarkets outsold the nonmodified variety by two to one.[17] But British consumers balked when Monsanto made it clear that if they wanted American soybeans in the future, their only choice would be the GM variety and the processed foods made from them. By Christmas 1998 sales of the modified tomato paste had dropped to almost zero. Over the next two years, seven large European supermarket chains joined forces to eliminate GM ingredients from their own brand products.[18]

Monsanto's corporate arrogance was like a starting pistol for Greenpeace. Its activists ranged onto the fields where GM test crops were growing and began pulling up the seedlings. Lord Peter Melchett, an organic farmer and former Labor minister who was head of Greenpeace's campaign against biotechnology, took his tractor onto a nearby field of genetically modified maize and mowed it to the ground. The public applauded. A jury acquitted Melchett and his cohorts, who admitted trespassing for what they saw as the greater good. The BBC's sixty-year-old radio soap, *The Archers,* "an everyday story of country folk," picked up the theme. Young Tommy Archer of the farming family was found not guilty of criminal damage after destroying a test crop of oil seed rape in one of his uncle's fields.

Almost overnight, British shoppers stopped buying GM foods. One day the supermarkets couldn't keep up with the demand for the cheaper GM tomato paste. The next day the little red cans sat on the shelves. By the summer of 1998, a poll showed that most British consumers wanted all products containing GM foods to be labeled. Supermarkets responded by racing to buy organic products. The American GM sales armies could not have done a worse job of selling GM to a wary public. One British frozen-food-chain manager recalled being told, "You are a backward European who doesn't like change. You should accept this is right for your customers."[19] Monsanto's public relations people even warned the antibiotech movement that Britons would have the new soy "whether they like it or not."

Meanwhile the probiotech British government pressed ahead with plans to test GM crops. The British scientific community already had its successes in gene technology. Watson and Crick had worked on the double helix at Cambridge University. Dolly the sheep was cloned in Britain. Tony Blair's Labor government hoped that the many products of biotechnology—pharmaceutical as well as agricultural—would spur the British economy into the twenty-first century. The government's chief scientific adviser, Sir Robert May, set the tone. "We have played a hugely disproportionate part in creating the underlying science; are we going to lose it like we lost things in the past?" [20] The government prepared for the imminent arrival into Europe of Monsanto's modified corn, soybean, and canola and tried to ignore the protests by a mixed bag of mystics, ecowarriors, and protectors of country life. Politicians, scientists, and agribusiness believed the outcry would not last long. To be anti-GM, they thought, was a passing eco-fad.

As it turned out, the roots of the movement were already deeper than they looked. All that was needed to change the political landscape was an incident or two, events to bring the movement into full bloom. That chance came in the summer of 1998. An organic farmer asked the government not to test GM maize so close to his organic corn for fear of genetic contamination. About the same time, a researcher in Scotland claimed that feeding GM potatoes to laboratory rats had slowed their growth and damaged their immune systems. Prince Charles joined the fray, emerging from the royal greenhouse, where he was famous for talking to his plants, to encourage the protest movement's daily progress.

For the next few months, the volatile mix of issues kept biotech foods in the news: organic farmers trying to avoid "contamination" from the new gene-altered crops; publicly funded scientists challenging the competence of their privately funded colleagues. The imprecise technology of genetic engineering drew critics, as did corporate dominance of (and now permanent interference in) the food chain. Little wonder that the public took fright, fearful of what dark secrets this technology held beneath the PR front of the agbiotech companies, and refused to buy the new foods.

The British anti-GM movement owes one of its early successes to

Guy Watson, owner of a large organic farm in Devon, who went to court in 1998 to stop government plans to plant a crop of genetically altered maize near his field of organic corn. Watson had been warned by the Soil Association, the biggest organic certification body in the United Kingdom, that his organic license would be withdrawn if they found evidence of cross-pollination, or gene flow, from the GM test crops to his own.

Watson's court case gave the Soil Association and other green groups an excellent platform. "Frankenstein Foods Threaten Organic Farmer" was the headline on the press release from Friends of the Earth about Watson's legal challenge. The Soil Association told journalists that "there could be no future for organic farming unless genetically engineered crop testing is brought under control and commercial planting prevented."[21] Instead of halting the field tests, the government sent a committee of experts down to Devon to investigate.

At issue was whether the two fields—Watson's corn and the government's GM test crop—were sufficiently far apart to prevent the GM maize pollen from being carried on the wind or transported by insects to Watson's crop. Studies had suggested that 200 meters would be a safe distance; the plans called for planting the GM maize at 275 meters. The official committee concluded that at 200 meters only one kernel in forty thousand of Watson's corn was vulnerable. Some committee members even thought that this proportion was too high.[22]

The Soil Association challenged the committee's figures. After calculating the speed of local winds on which the GM pollen might be borne, it concluded that the risk was much greater. One kernel in ninety-three, or about five kernels in each cob, were likely to be pollinated by the GM maize. And that calculation did not include the possibility of bees carrying the pollen even further; there were twenty beehives on the border of the trial site.

The test went ahead, but before the GM corn was old enough to produce pollen, activists went into the field one night and ripped the plants out of the ground. The action was a great propaganda victory for the antibiotech forces and the threat of gene flow or contamination from GM crops became an increasingly successful argument, not only in Europe but elsewhere. Scant attention was paid at the time to the actual facts of Watson's case. He had no plans to replant seed from his corn harvest, so in reality any gene flow from the test crop would not have been a problem. But to reveal this at the time would have spoiled the story. If Watson had caused an event, the even bigger row over the GM potato transformed by a gene from the snowdrop brought the issue front and center.

Each year, after winter's gloom, the romantic snowdrop celebrates the first hint of spring but, like many flowering plants, the snowdrop is not as innocent as it looks. Inside that delicate exterior lurks a nasty poison, a natural self-defense system that helps to protect the snowdrop against preying insects and pests. A drop of this toxin, from the lectin family, will kill an insect by attacking the tissue of its digestive tract and degrading its immune system. Even mammals can be at risk. Lectins found in red kidney beans can produce marked intestinal damage to humans, with severe diarrhea, unless the beans are properly soaked, rinsed, and thoroughly cooked.

Scientists in the new biotech laboratories quickly spotted the potential of the lectin gene. If it could be inserted into a major crop, such as corn or potatoes, it could enhance the plant's resistance to pests and reduce the need for spraying with chemicals. In the mid-1990s, one laboratory in England had already started such experiments with potatoes, but scientists wanted to know whether a lectin gene inserted into a food crop would be harmful to humans.

British government scientists were concerned that work being done on the risk posed by the transfer of alien genes was not keeping pace with the number of new products being tested. The number of tests in the United Kingdom was tiny compared with the number in the United States, where between 1989 and 1997, fifty-six transgenic plant species had been tested—including 450 trials of new genetically modified potatoes. Only twelve transgenic crops were being tested in the United Kingdom during the same period, but many scientists feared that the U.S. tests were not as rigorous as those in Britain or Europe.

At the government-funded Rowett Institute in Scotland, one of the oldest nutrition research centers in the world, scientists were looking at ways to overhaul and update the system of regulating new crops. In particular, the institute's researchers were trying to find methods of screening new GM plants for toxins and allergens. As in the United States, the current guidelines required companies to produce their own safety data, based on standards agreed on many years before the advent of genetic engineering. Most companies seeking approval for the release of GM crops in Europe had already obtained approval by the FDA in the United States, but that didn't seem to help. Alan McHughen, a researcher from Canada, had developed a GM variety of linseed, named Triffid, that had passed FDA and USDA inspection for marketing. However, when he sought British marketing approval, U.K. officials made it clear that they didn't put much credence in the U.S. regulations. The officials suggested that McHughen might omit the fact that Triffid had passed U.S. scrutiny, implying that it might prejudice his case.[23]

Rowett scientists had some fundamental differences with the FDA rules. They thought that methods for testing chronic and acute toxicity offered a "poor screen" for the more subtle effects of transgenic plants. "Existing assessments rely heavily on comparing the similarity of the transgenic protein with known allergens. . . . However, this assumes that . . . all allergens are known." This was not a "comfortable assumption," they stressed.

The institute viewed the FDA's concept of substantial equivalence—that a new tomato is basically the same as an old tomato—as no more than a "useful framework" for assessing safety. Routine analytical procedures for old, nontransgenic plants would be unlikely to detect new toxins in transgenic plants, they argued.[24] Rowett scientists not only worried about the overall philosophy, they also questioned the thoroughness of some of the tests that were routinely required by the Americans.

One of the key issues in the safety of GM foods was whether the human digestive tract could absorb the new protein made by inserting the alien gene. The FDA tests did not distinguish between the digestive capacity of healthy adults and that of the very young, the elderly, or those unable to produce the right amount of stomach acid. Rowett scientists also objected to the "accepted" American practice of testing proteins from the new gene.

For example, when Monsanto had applied to the U.S. Environmental Protection Agency to market its new transgenic potato with a Bt gene for protection against pests, the company had been required to show that the Bt toxin would not harm a variety of friendly insects, such as bees and ladybirds. But instead of extracting the toxin from its transgenic Bt potato, the company scientists produced the toxin from a Bt gene inserted into *E. coli* bacteria, then submitted it as "substantially similar" to the actual toxin in the new potato. The reason was cost. Extracting the toxin from the GM potato was more expensive because the toxin is produced at low levels. The EPA approved the test and such methods became standard practice.

The Rowett scientists argued that proteins can undergo different modifications in different hosts. The Rowett team called the bacterium test "unsound" and warned that it could "lead to premature conclusions about safety."[25] The issue was important because many of the genes being considered to provide insect resistance in transgenic crops disrupted the pest's digestive system as well as the proper working of its immune and hormonal systems.

A Rowett researcher named Arpad Pusztai was given government funds to look at new safety tests for genes that produce lectins in experimental transgenic potato plants. Pusztai, a Hungarian who had come to Britain as a refugee during the Cold War, was a world authority on lectins. Although a believer in biotechnology, Pusztai, like his colleagues at the Rowett, was convinced that new food plants with a gene that produced lectins needed to be tested more rigorously than the Bt potatoes from the United States had been.

Pusztai carefully extracted lectin from the transgenic potato, not from a gene in a bacterium as in America. Then he set up three experiments in which he fed rats potatoes laced with lectin, or containing inserted genes from a snowdrop or a bean that produced lectins. His previous experiments told him that pure snowdrop lectin was harmless to mammals, but he wanted to check again to see if that finding held true when the lectin was produced by a snowdrop gene in a transgenic food. One group of rats was fed ordinary potatoes laced with lectin. A second ate potatoes that had been genetically engineered with a gene to produce lectin, and a third control group dined on ordinary potatoes.

At the beginning of his third year of research—1998—Pusztai became concerned by preliminary data showing that rats fed the transgenic potatoes showed a slight retardation of growth, plus a degrading of the immune system. His research was not complete when the British media, eager for any stories about "Frankenfoods," heard of his experiments and asked him to appear in a TV program about GM foods.[26]

Pusztai agreed, with the approval of the Rowett Institute. The TV presenter asked the obvious question: "So, if genetically altered foods can affect rats in this way, could they possibly have long-term effects on humans too?" Pusztai was cornered. He could have replied that it was far too early for such a judgment and that he would not want to comment until his work had been peer-reviewed. Instead he said that he would not eat genetically modified foods if he could help it, until there was more evidence about their safety. And then he added a bombshell: "It's very unfair to use our fellow citizens as guinea pigs."

The media pounced on his words. Phones at the Rowett Institute rang off the hook with calls from European government officials, green groups, and biotech industry representatives from all over the world. Monsanto, which was funding some research at the institute but not Pusztai's, wanted to send a team of experts on the next plane from the United States to find out exactly what Pusztai had discovered. One of Pusztai's colleagues at the institute wondered if his research grant from Monsanto might be in jeopardy because of the company's obvious displeasure at all the adverse publicity. Antibiotech groups declared that here was another example of how publicly funded research groups—universities and institutes like the Rowett—were being dictated to by multinationals, in a trend that a Labor member of Parliament, Alan Simpson, called "one of the most corrupting influences of our time."

Much confusion surrounded exactly what Pusztai's preliminary results had suggested. Even the institute staff was not sure because Pusztai had not shown them his data. In Pusztai's defense—and their own—the institute's top staff emphasized the provisional nature of the results, which, they pointed out, had not been peer-reviewed. They added that if Puzstai had found some abnormalities, it was not really surprising. Lectin was a poison: if you feed rats only potatoes laced with lectin, they don't die but they may get sick. The point of the tests was not to find out if eating lectin was safe for mammals, it was to develop new methods of testing the effects on the potato plant's structure of alien lectin-producing genes.

In their panic and confusion, the institute's staff decided that the best way out of the mess was to hold an internal inquiry during which Rowett scientists would be forbidden from speaking with the media on the subject of snowdrops and potatoes. But they also decided that Pusztai would be retired immediately. (He had been kept on to do the lectin work.) The headlines, of course, declared that a GM scientist had been muzzled over scary results from Frankenfoods.

Puzstai found support from a handful of scientists who thought that he had been badly treated and that his brilliant career in lectins had been brought to an end unfairly. But other colleagues thought he had been foolish to release even a hint of unpublished, non-peer-reviewed data—especially data clearly destined to be controversial. The Institute's internal inquiry concluded that the thrust of Pusztai's work was perfectly valid—to develop new testing methods—but that his experiments were "too crude and preliminary to justify any claims for novel findings of either lectin-related or general biotechnological significance."[27]

Amid the confusion, Pusztai came to believe that he had found something wrong, not with the lectin per se but with the "process" of genetic engineering. Greenpeace admitted that the cause of the rats' maladies was still unclear but nevertheless called for "an immediate total ban on GMO food," and an end to "using millions of people as guinea pigs."[28] Without waiting for further evidence, Greenpeace stated, "For all we know, they [the maladies] might have been caused by the virus used to transfer the alien DNA to the potatoes." It continued, "This is the same virus used in Monsanto's Roundup Ready soy that is available in markets around the world." The author was Benedikt Härlin, then Greenpeace's genetic engineering coordinator. His reference was to the cauliflower mosaic virus and its 35S promoter. The implication was that if Pusztai's rats were harmed by eating GM potatoes, they might also be harmed by eating other foods.

After he retired, Pusztai also focused on the 35S promoter as a possible culprit. The 35S had been used to switch on the lectin gene in the British experimental potato. Pusztai wondered if the problem with the rats' digestive tracts might have been the result of something the 35S did once it was inside the potato. But he did not know and, as he pointed out, there had been no experiments to find out.

A special peer review was called by members of the Royal Society, an exclusive club of 1,200 British and Commonwealth fellows. A pillar of the British scientific establishment, the society included many of the scientists who had pioneered biotechnology. They quickly declared that Pusztai's experiments were "flawed in many respects of design, execution, and analysis." Whatever had caused the ill effects in the rats, it was probably not the process of the genetic engineering, but more likely the effects of starvation; potatoes don't contain enough protein to keep a rat going. Or perhaps they occurred because the potatoes weren't cooked; raw potatoes contain toxins. Pusztai's case was not helped by the discovery that the modified potatoes in his experiment contained 20 percent less protein than normal potatoes.

Green groups believed that the Royal Society had criticized Pusztai because it represented the British scientific establishment, which generally supported what they saw as biotechnology's vast promise. The environmentalists still noted that the society had not concluded that GM foods were safe. But scientists can't do that. "Although we have no evidence of harmful effects from genetic modification, this of course does not mean that harmful effects can be categorically ruled out," concluded the Royal Society. The absence of evidence is not evidence either of the risk or the safety of a new food. Only the day before, the British Medical Association, representing British doctors, had called for a moratorium on the planting of GM crops because of the uncertainty over their long-term effects on humans.

If the intention of the Royal Society had been to protect the biotech industry in Britain, it was not successful. The Pusztai affair would not go away. In October his research was published in an amended form in *The Lancet,* Britain's oldest medical journal. The paper, more circumscribed and cautious than Pusztai's public statements, restated that the GM potato diet had affected the intestines and stomach lining of the rats in various ways, but there was no reference to adverse effects to the immune system.[29] Pusztai and his colleague, Stanley Ewan, wrote that the abnormalities could be attributed to the snowdrop gene, or a fragment of the gene—the promoter, perhaps— in the DNA cassette. But the paper settled nothing. Rather, as *The Lancet* editor Richard Horton wrote, it "provides a report that deserves further attention."[30]

Some of the country's senior scientists thought that Pusztai's work should not have been published at all—and some apparently even tried to stop *The Lancet* from publishing it. Horton had submitted Pusztai's work to six reviewers—twice the normal number. A majority had agreed that it should be published, but Horton came under intense pressure to drop the paper. He was telephoned by a member of the Royal Society and, he says, threatened with loss of his job if the publication went ahead.[31]

Horton was especially careful in an accompanying editorial to present the findings as "preliminary and non-generalisable" and to write that publication was "not a vindication of Pusztai's earlier claims."[32] In addition *The Lancet* published a commentary by two Dutch researchers who wrote that Pusztai's experiments "were incomplete, included too few animals per diet group, and lacked controls. . . . Therefore the results are difficult to interpret and do not allow the conclusion that the genetic modification of potatoes accounts for adverse effects in animals."[33]

The Royal Society claimed Puzstai's paper confirmed its original judgment that the experiments were flawed. The popular media left the distinct impression that more had been at stake than the reputation of the unfortunate Pusztai, the Rowett Institute, or the editor of *The Lancet.* Big guns like the Royal Society were not brought out for such matters unless there were big stakes. The average reader was once again left in a state of confusion, caught between the forces of high commerce and low politics. Whether the 35S promoter was a risky tool in the hands of the bioengineers was a question that would be raised again by Ho and other scientists who continued to oppose the technology.

Meanwhile in America Monsanto saw the European market for its GM crops slipping away. At the company's St. Louis, Missouri, headquarters, executives prepared a counterattack. They looked for allies among the European biotech companies, such as the Swiss Novartis, the British AstraZeneca, and the German AgrEvo. Not surprisingly, they were turned down. The European companies had warned Monsanto about the potential of the opposition in Europe, and Monsanto had ignored them. When the European companies had sought a share of the U.S. biotech market, Monsanto had blocked them through broad patents, warding off any patent challenges with lawsuits. Monsanto also lived with a public image that had been tarnished for years. From 1935 to 1977, Monsanto was the sole maker of PCBs—the polychlorinated biphenyls (used in electronic transformers, pesticides, and lubricating oils) that had turned out to be carcinogenic. During the Vietnam War, Monsanto was among the companies that made Agent Orange, the herbicide sprayed on Vietnamese jungles and later blamed for cancers and birth defects. The "wicked chemical giant" of the past, now returned under a new banner, was again having trouble with its public image, especially in Europe. The latest news from the American biotech labs was not helping.

In the spring of 1998 the U.S. Department of Agriculture and a Mississippi cotton seed company named Delta and Pine Land had announced a new invention for plant breeders—a means of making plant seeds sterile. Here, it seemed, was the ultimate gadget allowing biotech companies to play God. Instead of taking seed from a harvest and planting it again next season, farmers would now plant a classic American product—a seed that could be used only once and then thrown away. Monsanto was trying to buy the cotton seed business of Delta and Pine Land, and all its inventions.

The international antibiotech group Rural Advancement Foundation International (RAFI) brilliantly dubbed the new technique "the Terminator" after the robotic killer played by Arnold Schwarzenegger. For those who worried about the effect of the Green Revolution and industrial agriculture on developing nations, the Terminator was seen as a threat to nearly a billion poor farmers who relied on saving next year's seed from their harvest.

The question for Monsanto was how to thrive abroad. Since 1995 the company had been run by Robert Shapiro, a Brooklyn-born lawyer and former urban affairs professor, who dreamed of turning the chemical giant into one of the futuristic "life science" companies. "This is an important moment in human history," Shapiro told the *New Yorker* in 1999. "The application of contemporary biological knowledge to issues like food and nutrition and human health has to occur. It has to occur for the same reason that things have occurred for the past ten millennia. People want to live better, and they will use the tools they have to do it. Biology is the best tool we have."

Shapiro's idea was that new corporate hybrids would employ the gene revolution in the service of humankind. They would provide new drugs and crops and even foods that take the place of drugs. At the same time, their products would help create a cleaner world by using fewer toxic chemicals. Within three years Shapiro had sold Monsanto's chemical operations and launched a multibillion-dollar buying spree of seed companies. By acquiring Agracetus, Asgrow, and Holden's, Monsanto became almost overnight the world's leading agbiotech company. Although the patent had run out on Monsanto's Roundup herbicide, a best seller for twenty-five years, the company was assured of continued profits with the production and patenting of seeds containing a new gene that made the crop plants resistant to the herbicide. A genetically altered corn plant resistant to pests was next. The company's stock price doubled. When opposition rose in Europe, Shapiro thought it was a storm that would blow over, a passing whim of activists who were going to be opposed to technological change wherever and whenever it came.

Inside the company, however, there was a more realistic assessment of the situation. Some executives had understood the power of the opposition from the beginning and believed that the protesters had to be met at least halfway.[34] When the company launched its advertising campaign to counter the protests, the division of views inside the executive suite was evident. One ad encouraged consumers to hear all opinions and gave the phone numbers and e-mail addresses of protest groups. Another claimed the moral high ground, asserting that biotechnology would feed starving millions.

The campaign was a spectacular failure, not only because of the mixed message but also because of Prince Charles. In a series of royal pronouncements, which happened to coincide with Monsanto's ad campaign, Charles plunged into the biotech war. "This kind of genetic engineering takes mankind into realms that belong to God, and to God alone. . . . I personally have no wish to eat anything produced by genetic modification, nor do I knowingly offer this sort of produce to my family or guests." GM crops, he said, were unnecessary and incompatible with agriculture that "proceeds in harmony with nature."

The antibiotech forces were overjoyed, of course, but Charles could not sustain the attack. His arguments soon moved into mystical musings; he argued on the BBC that the best guide to what is right for the planet is not rational thought but "a wisdom of the heart, a faint memory of a distant harmony, rustling like a breeze through the leaves." Such opinions did not travel well in some parts of the developing world. Chengal Reddy, president of the Andhra Pradesh Farmers Association, complained, "It is like someone telling me when some disease like malaria or bronchial asthma affects me that I'm not supposed to use modern medicines." [35]

Charles also had his critics at home. Richard Dawkins, the zoologist and writer, said it was "no use appealing to 'nature' or to 'instinct' when trying to decide which type of agriculture to pursue. He was concerned that the prince had "an exaggerated idea of the 'naturalness' of 'traditional' or organic agriculture. Agriculture has always been 'unnatural.'" [36] Nature is cruel, said Dawkins—"red in tooth and claw," as Tennyson had observed. "It may sound paradoxical, but if we want to sustain the planet into the future, the first thing we must do is to stop taking advice from Nature. Nature is a short-term Darwinian profiteer. Darwin himself said it: 'What a book a devil's chaplain might write on the clumsy, wasteful, blundering, low and horridly cruel works of nature.' "

Prince Charles was on one flank of the antibiotech forces and on the other were anarchists and ideological scientists like Mae-Wan Ho. In the middle were millions of trade unionists, religious groups, Greenpeace, the World Development Movement, Friends of the Earth, and Christian Aid.[37] The Church of England was forced to speak up. On the question of humans "playing God," the church pointed out that "much technology and most medicine is based on human intervention in the natural processes. Human beings are themselves part of nature, creatures within creation." Therefore, human discovery and invention could be seen as the exercise of God-given powers of mind and reason. To have the power to invent was what it meant to be "in the image of God."

In a reference to "Frankenfoods," the church, which owns more than 123,000 acres of agricultural land, warned against being "unduly influenced by slogan words" and, sitting itself squarely in the middle of the debate, pronounced that it "would be unwise, either to ban GMOs from foods, or to fail to keep their use under scrutiny."[38] The Catholic Church appealed for honesty and openness about the technology. The Vatican, which was adamantly against the cloning of humans in all its forms, gave a cautious yes to tinkering with plants and animals. "We cannot agree with the position of some groups that say it is against the will of God to meddle with the genetic makeup of plants and animals.[39]

By the fall of 1999, GM foods had overtaken "mad cow" disease as the British public's biggest food safety concern, with a fifth of the people polled saying they would never knowingly eat or feed their family anything containing gene-altered food. Monsanto's strategy had hopelessly backfired. Eighty-five percent of Britons surveyed thought they were being denied access to all the facts. At the same time, sales of organic foods were booming. With a twentyfold increase in sales in three years, organic products now made up 3 to 4 percent of all food sold in supermarkets. Farmland acreage devoted to organic crops in Britain increased fivefold in 1998, but still only accounted for 1.5 percent of the total.

These trends were in opposition to Tony Blair's government, which had gone out of its way to support the technology, anxious to keep Britain in the vanguard of the revolution. The ultimate embarrassment for Blair was a notice that went up in the House of Commons dining room announcing that the use of GM foods would be "avoided wherever possible." The decision had been taken unilaterally by the Commons refreshment department because of general unease about the new foods. Blair stuck to his guns. "All I say to people is: just keep an open mind and let us proceed according to genuine scientific evidence."

By the end of 1999 the greens were declared the winners of the first round against Monsanto. The company's fall was hard—from market leader to stock price slump amid rumors that the corporation was going to divest some of its holdings. The cranky minority, as Shapiro had once seen the protesters, had turned out to be much more powerful than he had imagined. Billions of dollars later, Shapiro's idea of the life science company with agriculture, pharmaceuticals, and nutrition all under one roof was coming unstuck.

Leaders in agbiotech were merging and consolidating. Monsanto agreed to merge with the pharmaceutical giant, Pharmacia & Upjohn. The combined company would be run from Pharmacia headquarters in Peapack, New Jersey, a decision that symbolically severed ties with St. Louis, where Monsanto, as a small chemical company, had started life at the turn of the century in the garage of John Queeny and his wife, Olga Monsanto. The *Wall Street Journal* said Shapiro's life science concept had become an "affliction" for the company. "The crop biotechnology half of the program has grown so controversial that

Monsanto has agreed to a deal that is likely not only to push biotech to the back burner, but also to cost Monsanto its independence. And investors are reacting harshly.[40]

In an extraordinary display of contrition, Shapiro went before the enemy (on a video screen) at the fall 1999 annual meeting of Greenpeace and apologized for his mistakes. "Our confidence in this technology and our enthusiasm for it has, I think, been widely seen—and understandably so—as condescension or indeed arrogance," he confessed. "Because we thought it was our job to persuade, too often we forgot to listen." Lord Melchett, then still Greenpeace's biotech leader, was supposed to debate Shapiro, but he didn't know quite how to respond. Melchett suggested that if Monsanto would renounce all biotech products and embrace organic farming, then Greenpeace would become the company's partner in finding a solution to the world food problem. "It was a bit like offering moral support to General Motors, if only the automobile maker would abandon the internal combustion engine in favor of the bicycle," observed Michael Specter in the *New Yorker*.[41]

Anatomy of a
Poisoned Butterfly

It's not an exaggeration to say more monarchs succumb to high-velocity collisions with car windshields than ever encounter corn pollen.

—Val Giddings, Biotechnology
Industry Organization, 1999

We worked very hard to make this a high-profile issue. . . . The question still remains, would this science have been done if the monarch wasn't such a beautiful butterfly?

—Margaret Mellon, Union of
Concerned Scientists, 2001

On the question of endangered species, most people cannot get worked up about the snail darter, the dwarf wedgemussel, or the oval pigtoe. Mention the possibility that the hairy rattleweed or the mat-forming quillwort is stressed out and a hiking club somewhere might be moved to protest. But suggest that the life of the common and beloved monarch butterfly might be at risk from human hands and a whole nation starts making posters.

The brilliance of an American summer would indeed be dimmed without the gaudy orange-and-black creatures dipping and diving in the meadows. The richness of an American fall would certainly be dulled without the monarch's wondrous southern migration, an incredible journey during which tens of millions of creatures fly eighty miles a day to spend the winter in a Mexican forest roost. So the news that big agriculture might be killing off the royalty of American insects had all the makings of a biodisaster.

In the spring of 1999, as the monarchs embarked on their return flight north, a young Cornell University entomologist named John Losey reported in the journal *Nature* that the monarch's future appeared to be endangered, not from urban sprawl or toxic waste but from eating the pollen of genetically modified corn. At the time twenty million acres of American farmland, representing a quarter of the U.S. corn crop, had been planted with seeds that included a toxin-producing gene from the common soil bacterium, *Bacillus thuringiensis* or Bt. The insect-poisoning power of Bt had been known for over a century; the first commercial spray was developed in Europe during World War II. Half a century later there were 182 Bt products registered by the EPA.[1]

Two other big crops—cotton and potatoes—had also been fitted out with the Bt gene. In corn the Bt toxin was designed primarily to kill the European corn borer, a caterpillar that destroys more than $1 billion worth of the crop each year. The toxin punctures the delicate membranes of the corn borer's digestive tract, causing it to wither and die.

Most of the monarchs born in the Midwest corn belt start life on a milkweed leaf in or around the edges of a farmer's land.[2] When the corn sheds its pollen during July and August, pollen grains containing the Bt toxin are blown by the wind onto milkweed leaves. From earlier studies, Losey knew that Bt toxin could harm butterflies and moths, and he wondered if the monarch larvae might also suffer.

In a no-frills experiment, he fed monarch larvae with Bt pollen in his laboratory at Cornell. If they showed signs of harm, he intended to do more research in the field. In his lab, he misted milkweed leaves with water and sprinkled on the Bt corn pollen to a density that looked like the pollen he had observed on the milkweed in a cornfield. He then placed five three-day-old monarch larvae—caterpillars no bigger than a raindrop—on each milkweed leaf and watched them feed. After four days, nearly half of the larvae were dead. Those that survived were half the weight of his control group feeding on milkweed leaves with no pollen. Larvae fed on leaves sprinkled with conventional hybrid corn pollen were still munching away, apparently no worse off.

Losey and his Cornell researchers were well aware of the uproar the deaths of even a few monarch caterpillars could cause. Bt crops—cotton, corn, and potatoes—were the green-friendly pride of the biotech industry. The idea was that farmers who planted them would use fewer pesticides because Bt would kill all the corn borers. One estimate suggested that of the $8.1 billion spent annually on insecticides worldwide, nearly $2.7 billion could be saved by substituting Bt biotech products.[3] Elsewhere around the world, other borers—the Asiatic, the southwestern, the corn earworm, the fall armyworm, and the black cutworm—tunnel into corn stalks. They are difficult to control with chemicals because they are hard to find inside the corn stalk and spray effectively. In America, corn farmers had mostly given up. The amount of pesticide being used against the corn borer had remained constant for years. With corn yields rising, it was cheaper not to bother with the little beast.

When Bt corn came along, it quickly became a popular way for farmers to manage the corn borer. The use of Bt varieties had increased dramatically since the first planting in 1996. After the opposition to GM crops in Europe, the industry was hoping for a breathing space, not more bad news. Short of some human health hazard, it was hard to think of a bigger propaganda setback than monarchs being killed by GM corn.

In America, people of all ages are so enamored of the monarch that some chart almost each flutter of its migration from Mexico to the Canadian border and back again. En route the monarchs mate in an elegant embrace and the females each lay up to four hundred eggs, usually on the underside of milkweed leaves for maximum protection. The monarch has its own website, Monarch Watch, started by Orley "Chip" Taylor, an entomologist from the University of Kansas. Every year monarch watchers faithfully record the first landing of the butterflies onto milkweed leaves, as well as the first and last hatch of the season. Milkweed is the monarch caterpillar's food and its home. The butterfly's egg is a creamy dome, something of a pearl for lepidopterists, who hail it as a natural wonder. One fan rhapsodized about a "priceless gem cut by a master craftsman . . . one of the most exquisite objects in nature."[4]

During the summer, when the butterflies are busy eating and reproducing, they live for up to six weeks. Those born in September instinctively forego the debilitating rituals of courtship and sex, saving their energy for the three-thousand-mile flight south. Their determination to make this arduous journey has kept even expert entomologists scratching their heads in wonder. How do the monarchs know when to begin the migration? How do they find their way? What proportion of monarchs survive the journey?[5] Such mysteries only add to the creature's enduring popularity. In fact, the monarch has become such a hot entomological property that some specialists bemoan the way it has upstaged other beautiful butterflies, such as the swallowtail, the clouded sulphur, the checkered white, and the handsome buckeye. The lepidopterists' website Butterflies.com includes an "Essential Butterfly Guide" subtitled, "There is more to life than the Monarch."

To test public reaction to their experiment, Losey and his co-researchers at Cornell first shared the results with colleagues. All were in favor of publication, says Losey. However, a senior entomology professor at Cornell, Anthony Shelton, warned the younger researcher, "You don't have a story here."[6] Professor Shelton, a believer in biotech generally, would become increasingly unhappy that Losey's experiment had been confined to a laboratory. The results, he would later complain, were "not pertinent to the real world."[7]

This criticism from a peer put Losey in a bind. As an assistant professor, aged thirty-seven, he was up for tenure in 2004. He needed to publish his work, but he also needed the firm support of his senior colleagues. Finally Losey decided to write the report and send it to science journals. "To have sat on the data and not publish would have been unprofessional," he said later. "And it would have been irresponsible not to take the results to our peers and the public." [8]

In search of a publisher, Losey's first stop was *Science* magazine, the journal of the American Association for the Advancement of Science. It was not interested. The next stop was the British weekly journal *Nature*, where Losey had previously published. Intrigued, its editors sent the report to two reviewers, who urged publication. After minor changes, the report—a mere seven paragraphs long—was published in March 1999 under the title "Transgenic Pollen Harms Monarch Larvae." Losey warned that Bt corn pollen could have "potentially profound implications for the conservation of monarch butterflies," [9] a message provocative enough to make the front pages in monarch-conscious newsrooms across the United States.

Headlines blared the results. "Butterfly Deaths Linked to Altered Corn," warned the *Boston Globe*. "Gene Spliced Corn Imperils Butterflies," said the *San Francisco Chronicle*. "Man-made Corn May Do More Harm than Good," declared the *Orlando Sentinel*. "Altered Corn Kills Butterflies," said the *Denver Post*. The *New York Times* put a picture of the monarch on its front page over a caption calling it "the Bambi of the insect world."

The antibiotech forces leapt into action. Greenpeace called for an immediate ban on the planting of Bt corn. Its volunteers dressed up as monarchs that collapsed as they were "felled by killer corn." The more reflective of the green activist groups, such as the Union of Concerned Scientists (UCS) and Environmental Defense, noted that Losey's results demonstrated how the government's insecticide regulatory department, the Environmental Protection Agency, had failed to address the real risks to the monarch—and other insects, for that matter—before allowing Bt corn into the environment. "One cannot help but wonder what other, perhaps less obvious, environmental impacts of genetically engineered crops have been missed by the EPA," said the UCS biotech director, Margaret Mellon.

Before Bt products came on the market, the EPA had examined public and company reports of the effects of the toxins on a variety of organisms that might be found close to Bt crops. These included birds, fish, honeybees, ladybugs, parasitic wasps, lacewings, springtails, aquatic invertebrates, and earthworms. The EPA had concluded there were "no reasonable adverse effects" to humans, the environment, or any organism that Bt was not supposed to kill.[10] Scientists knew that Bt toxins could be harmful to the larvae of lepidoptera, but the EPA had looked primarily at exposure of larvae from eating leaf tissue, not pollen. The agency had considered the possibility of pollen drift from Bt cornfields but concluded that the pollen was not toxic, even at relatively high doses.[11] The EPA did not specifically require tests for effects on the monarch larvae because it did not believe monarchs were likely to be present in or around cornfields.[12]

Losey had identified a gap in the research, and the green groups rushed through it. Dr. Mellon of the Union of Concerned Scientists recalled, "We worked very hard to make this a high-profile issue because without media attention we knew nothing would be done. We saw the findings as an illustration of how superficial risk assessment [for genetically modified foods] was. . . . The question still remains, would this science have been done if the monarch wasn't such a beautiful butterfly?"[13]

Dr. Mellon's group reminded the public that Losey's research could spell trouble for many other lepidoptera—moths and butterflies feeding in the vicinity of Bt corn fields. On the U.S. Endangered Species List were nineteen threatened butterflies and moths. And only the year before, Swiss scientists had reported laboratory results showing that Bt corn was harmful to green lacewings, beneficial insects that feed on pests, including the European corn borer. In the Swiss study, lacewings fed corn borers that had eaten Bt corn had a higher death rate and delayed development.[14] In their submission to the EPA, the companies had persuaded the government agency that lacewings would be immune to Bt toxins. Incredibly, the companies themselves had never assessed the risk to America's favorite butterfly—or if they had, the results had not been made public.

Other groups, including Chip Taylor's Monarch Watch website, pointed out that Bt corn wasn't the only new biotech crop that threatened the monarch. The development of genetically modified herbicide-resistant crops had allowed farmers to use broad-spectrum weed killers to rid fields of weeds, including the monarch's favorite milkweed. And the monarchs' winter habitat in Mexico, so well camouflaged that it was only discovered by American researchers in 1975, was now threatened by modern development. The monarchs might be in real trouble.

As spring gave way to summer in 1999, Losey's study came under scathing attack from Cornell's Professor Shelton, who was increasingly agitated by the media fuss the report had generated. At one point he suggested that Losey was no better than a rumormonger. In Congressional testimony, he invoked the character Rumor in Shakespeare's *Henry IV, Part Two*. It was Rumor who falsely announced that Hotspur had triumphed at the Battle of Shrewsbury when, in fact, Hotspur had been not only defeated but killed. The play describes the chaotic events resulting from the spread of the rumor.

In Shelton's view, the media reporting of Losey's *Nature* article had produced an equivalent and equally unwelcome modern drama. "Was this reaction justified based on what can only be considered a preliminary laboratory study? Absolutely not!" said Shelton. "We cannot afford to be an ignorant society on these important new technologies and fall victim to false Rumor."[15] Other scientists were also upset. A follow-up letter in *Nature* from a British biologist counseled the need for more "scientific rigor" in the presentation of information "to ensure that it is not misrepresented" and warned that "preliminary observations should not be overinterpreted."

But Professor Shelton became Losey's main critic. In a "Cornell University Press Release," Shelton, with University of Adelaide researcher Richard Roush, attacked Losey's experiment. "If I went to a movie and bought a hundred pounds of salted popcorn, because I like salted popcorn, and then I ate those salted popcorn all at once, I'd probably die," Shelton was quoted as saying. "Eating that much salted popcorn simply is not a real-world situation, but if I died, it may be reported that salted popcorn was lethal. The same thing holds true for monarch butterflies and pollen. Scientists have a duty to be incredibly responsible for developing realistic studies. Scientists need to make assessments that are pertinent to the real world. . . . Few entomologists or weed scientists familiar with the butterflies or corn production give credence to the *Nature* article."[16]

The biotech industry also attacked the artificial nature of Losey's data, focusing especially on the fact that the monarch larvae had eaten the pollen in the controlled environment of Losey's laboratory. It was "highly likely that in the natural setting, outside the laboratory, most monarch larvae would never encounter any significant amounts of corn pollen," declared an industry organization press release.

But this was a bluff. The industry didn't really know what happened to monarch larvae feeding on milkweed leaves in or near Bt cornfields because the research had not been done. A big company involved in Bt crops, like Monsanto, might have been expected to carry out relevant studies, but there was nothing in the public record. Even so, Monsanto joined in the attack on Losey, asserting that the exposure of milkweed to corn pollen is "very low because only a small portion of milkweed grows in close enough to cornfields for exposure to corn pollen."[17] Again, there was no public scientific evidence to back up such a statement.

The heavy hand of the biotech industry PR machine would offer the information that more monarch butterflies were killed colliding with car windscreens "than ever encounter corn pollen." In fact, the biggest hazard for monarch larvae is to be eaten by other insects. Fewer than 10 percent survive to adulthood.[18] The industry could have added that the mowing of highways, ditches, and pastures, not to mention urban sprawl and chemical spraying, threatened not only the monarch larva's milkweed home but also the adult butterfly. At least there was some public evidence for these common dangers.

Such corporate petulance was hardly justified, given Losey's efforts at evenhandedness. After *Nature* had accepted his report, but before it was published, Losey contacted Monsanto and Novartis, the two companies involved in Bt corn, to let them know of the upcoming article. "I wanted to be aboveboard and not blindside them," he said.[19] Monsanto and Novartis hastily dispatched staff scientists to visit Losey and argue that the science was not "robust enough" for the generalizations he had made. "They wanted more detail," said Losey. "There was clearly a feeling that we should not publish at this time; that we should wait until we got more data."[20]

After the publication of his report, Losey was quoted in Monsanto's *PR Newswire* as one of "several academic experts [who] have urged caution when interpreting the results" of his own study. "While [our study] raises an important issue, it would be inappropriate to draw any conclusions about the risk to monarch populations in the field based solely on these initial results," Losey said. It was a distinctly different tone from his first conclusion in *Nature* that "these results have potentially profound implications for the conservation of monarch butterflies." Losey had been taken to the woodshed, but if he was humbled, he was not contrite. He had always readily acknowledged that his experiment was only a preliminary study needing extra research. He made this point over and over again in a flood of media interviews.

The young Cornell researcher took the industry's flak, but in fact he had not been the first to report that Bt corn pollen had proved lethal to monarch larvae. Another team of researchers from Iowa State, John Obrycki and Laura Jesse, had also fed monarch larvae with milkweed sprinkled with corn pollen and some of their larvae had died.[21] In their three-year study, they had attempted to recreate at least some of the field conditions. They put potted milkweed plants in cornfields during the corn's pollen shed, then took the plants back to the lab and put larvae to feed on the leaves. Some had died. They had finished their study and reported their results to colleagues and the industry before publication of Losey's *Nature* paper, but their results were not published until a year after Losey's, at which point they generated even more alarm. The Iowa team claimed to have the first evidence that transgenic Bt corn *naturally* deposited on milkweed in a cornfield causes significant mortality.[22]

The industry complained that the Iowa work was not a realistic field test either. Some researchers agreed. Kevin Steffy, an entomologist with the University of Illinois Extension Service, suggested that it was only a "modified field study." In his view, the media had overreacted to both studies. "Quite frankly, I'm getting tired of the press making an issue out of scientific findings that don't describe the real world terribly well. I am also dismayed by some of the sweeping conclusions the authors [make]. . . . I will not argue with the assertion that the potential limitations of Bt corn and other transgenic crops need to be studied. . . . If scientific evidence reveals negative impacts of transgenic crops, then let the chips fall where they may. But let's be very careful about interpretations of scientific studies."[23]

Whatever anyone thought of Losey's or the Iowa study, the researchers had clearly identified a hazard for the monarch butterfly. The biotech companies moved swiftly to control the damage, calling for more research and putting up funds to convene an inclusionary process and have third parties develop the data.[24] Several academics, including Losey and Chip Taylor of Monarch Watch, were invited to carry out studies funded 60 percent by industry with the rest of the funding coming from government grants and other sources. Losey said he wanted to research whether monarchs would avoid pollen-dusted leaves, but the industry was not interested in his inquiry, so he bowed out.[25] The industry said Losey wanted to "take the research in his lab in a different direction" from the one they had chosen.[26]

The idea that the companies were now funding public research into a biotech hazard was a significant departure from the cozy relationship they had enjoyed thus far with the FDA and the EPA. Both agencies had required only voluntary company research by their own in-house scientists—and the results were kept confidential whenever the company claimed trade secrets. In agreeing to this new, more open way of doing things, the biotech industry was hardly running up a white flag, however. Company scientists and many outside researchers still believed that they would be proved right—that Bt corn was not a hazard to the monarch butterfly.

In their view, Losey's laboratory experiments had been amplified by media so caught up in the high drama of the caterpillar deaths that they failed to ask the larger questions. Had the media done so, the companies were of the opinion that the risk of Bt corn to the monarch would never have entered antibiotech folklore.

The ecological risk to an insect depends on the possibility of exposure to a known poison and then on the effectiveness of that dose. Losey's brief laboratory experiment had no real information on either of these criteria. And there were no studies on the chances of the tiny tiger-striped mite of a monarch larva being hatched on a milkweed leaf close enough to a Bt cornfield for exposure. No real research had been done on whether the hatching of the monarch larva was likely to coincide with pollen shed from the corn.

The larval stage lasts between twelve and sixteen days; the corn plant sheds pollen for about seven to ten days. Even if these two events overlap, there is still a question whether enough pollen would settle on the milkweed leaf—and remain on the leaf during the time the larvae are feeding—to have a harmful effect. The pollen density on the milkweed leaves used in Losey's experiment was a rough estimate, "set to visually match the densities on milkweed leaves collected from cornfields." In other words, he eyeballed it. When he was "gently tapping a spatula of pollen over milkweed leaves that had been lightly misted with water" (his method of putting pollen on the leaves in the lab), he kept going until the leaf in the lab looked roughly like the leaves he had seen in the cornfield.

In Losey's lab, the larvae were force-fed for four days on leaves constantly covered in pollen. In the field, the number of grains is not likely to remain constant. Pollen blown onto a leaf by the wind is also blown off again. Pollen can also be washed off by a shower of rain, or even a heavy dew. In addition, not all pollen from Bt corn contains the same amount of Bt toxin. At the time of Losey's study, there were several different types of Bt corn made by agbiotech seed companies, including Dow AgroSciences, Monsanto, and Novartis. Each type was slightly different.

Losey chose pollen from a Bt corn known as Bt11, made by Novartis, because that was the corn pollen readily available at Cornell at the time he began his study. The Bt gene in Novartis's Bt11 (and Monsanto's Mon810) was controlled by the controversial 35S promoter gene from the cauliflower mosaic virus. The 35S turned on the Bt gene in every part of the corn plant, including the pollen.

But another type of Bt corn had distinctly different characteristics—especially for a monarch caterpillar. Known to its Novartis inventors as Event 176 (biotech engineers tend to call each new product an "event") and sold under the trade name Knockout, this type of Bt corn was modified to produce a much larger dose of toxin in the pollen. Although the corn borer does its damage to the stalk, pollen grains tend to collect at the base of the cob, where the corn borer also feeds. Scientists created extra toxin by adding a special poison-boosting promoter that increased the poison in the pollen grains.

Event 176 was a much more powerful exterminator—containing

up to ten times more toxin in the pollen than the other types.[27] Two key questions therefore needed to be answered by the industry's new studies: What were the chances that monarch caterpillars might be exposed to such pollen? And how much pollen is required to have a toxic effect?

To limit the damage done by Losey's article, the industry rushed to complete its research during the 1999 growing season. Field studies had to be in place in late July and early August, when the corn shed its pollen, otherwise the work would have to wait for another year. By the fall of 1999, preliminary papers were ready. In November, the scientists reported to a symposium in Chicago organized by the industry to be as open as possible. Outside academics not involved in the studies were present, as were environmentalists who had been critical of biotechnology, and the media.

Not one study reported the kind of mortality in monarchs that Losey had found. But the impact of that news was blurred. In their eagerness to be first to define the results, the industry's public relations squad had called a few selected newspapers the day before with their interpretation that Bt corn presented little risk to monarchs. On the day of the symposium, several major newspapers, including the *Los Angeles Times,* the *Chicago Tribune,* and the *St. Louis Post-Dispatch,* all reported that the symposium would conclude that Bt pollen posed little risk.

But the industry's spin operation backfired. In fact, a number of scientists at the Chicago meeting found cause for continued alarm. Of three Bt corn "events," the strongest—Novartis's Knockout, or Event 176—produced pollen that was found to be highly toxic to monarch caterpillars. Pollen from the two others was found to be less toxic. The implication was that only Event 176 would kill the larvae. But some researchers pointed out that Bt toxins in pollen lose their potency after being stored for a week and that the study had not made allowances for the pollen's having been stored.

Several researchers at the symposium protested the industry's premature interpretation of the papers. They said that there was not enough clear evidence to come to the industry's conclusion; they demanded longer studies. Even if two types of Bt corn did not kill monarch larvae, they might be impairing growth in some unseen way.

Rebecca Goldburg of Environmental Defense summed up the feeling of the green groups: "It appears that such questions will only be addressed if there is funding for research on Bt corn pollen and butterflies independent of the industry." Such funding would "also help to insulate researchers from the pressures of commercial interest."[28] Opinions on the meeting were polarized. One researcher called it a "travesty"; another said press complaints about the industry spin were undeserved.

Obviously, the industry still had a fight on its hands. The initial five-year registration of Bt corn—the first approval of a GM organism by the EPA—was running out. Losey's paper and the inconclusive results of the Chicago meeting now prompted the EPA to take a harder look at the data. The agency asked all seed companies producing Bt corn to submit new research about the toxicity of corn pollen, showing the level at which monarchs would be exposed and the potential impact of that level on monarch populations. The studies had to be completed by the spring of 2001 so that the EPA could make a decision on the future of Bt corn by the fall. Farmers had to know by then in order to purchase seed in time for the 2002 growing season.

The biotech industry and the U.S. Department of Agriculture each put up one hundred thousand dollars to fund twenty-six academic and government scientists, including John Losey. They would write six papers on all aspects of Bt corn's possible threat to the monarch butterfly. Their work would be submitted to the Proceedings of the National Academy of Sciences, which requires two outside reviewers for each paper. Industry researchers were confident that the outcome this time would clear Bt corn, but they couldn't be certain. In the meantime, green groups unearthed yet another surprise hazard.

By the end of 1999, the intense media coverage of Dr. Pusztai's snowdrop-laced potatoes in Britain and then Losey's monarch butterflies in America had spurred the antibiotech forces to a new offensive. In the United States the greens launched an ad campaign in major newspapers about the "gravest moral, social, and ecological crisis in history." The ads asked, "Who plays God in the 21st century?" and declared GM food was "unlabeled, untested . . . and you're eating it."

Funds flowed into the green group coffers from foundations flush from a high-flying stock market. A specific target of the regenerated campaign was Bt crops: if they could harm butterflies, what about people? Friends of the Earth canvassed members with a new campaign. "How safe is the food you eat?" asked the fund-raising letter. "If deadly toxins that kill butterflies are being introduced into our food supply, what effect are these toxins having on you and your family? . . . The scary answer is that no one really knows." Other groups appealed for funds with envelopes adorned with monarch butterflies.

At the same time, the antibiotech campaign turned to other issues. The organizers began to target power rather than science, taking on patents, corporate power, globalization, world hunger, poverty in the Third World, the issue of sustainable agriculture, and the legacy of the Green Revolution. Protests against the activities of greedy capitalists would make more headlines than the sticky scientific details of transgenic plants. The World Trade Organization became as much a target as Monsanto. Protesters shouting "No, no, to GMO" rioted in Seattle in December 1999. Food companies took fright. Heinz and Gerber removed GM ingredients from baby food. Europe and Japan suspended reviews of Bt corn. Japanese brewers said they would not use GM grain in their beer.

Scientists who had been involved in biotech research from the beginning, and who worried that companies like Monsanto had skipped too easily through the regulatory hoop, were quietly pleased to see the antibiotech forces creating such a fuss. More research was needed, they knew. The louder the protests, the better the chances of a mid-course correction and more funds for their inquiries. In the new offensive by the antibiotech campaign, however, there were always opportunities for an adventurous activist to rattle the scientific foundations of GM.

One of the oddities among the Bt corn products was a type traded as StarLink, made by the giant European seed company Aventis. All types of Bt corn produce crystalline proteins made from the Bt gene and known by the prefix *Cry-*, for *crystalline*. For example, the Bt11 type used by Losey produced the Cry1Ab protein and StarLink produced the Cry9c protein. When a company applied to the EPA for approval of a Bt corn, the Cry protein was put through a rather crude human allergy test. Bt proteins were placed in an acid solution that mimicked human stomach fluids, where usually they broke down readily into harmless amino acids. Researchers concluded that there was no likelihood of the protein's staying around long enough to cause an allergic reaction in humans.

The stubborn Cry9c was different. It remained stable for more than an hour in the acid solution, behaving much like most known food allergens. The delay was long enough to give the body time to react. StarLink was not poisonous to rats, nor was its biochemical structure similar to that of the majority of food allergens, so its stability in stomach acid did not mean it would automatically cause an allergic reaction, even in sensitive human stomachs. But when the makers, Aventis, applied for a license in 1997, the EPA, unsure of its safety for human consumption, approved StarLink for animal feed only. The approval implied that StarLink corn could somehow be kept totally separate in the U.S. grain system from products intended for humans, a zero contamination standard just like the standard the Europeans wanted for any U.S. grain imports.

After speaking with corn farmers about the difficulty of separating one corn type from another, Larry Bohlen, an activist with Friends of the Earth, decided to test food products to see if any StarLink had crept into the food grains. One night in the summer of 2000, Bohlen went to his local supermarket in Silver Spring, Maryland, bought a grocery cart of corn products—breakfast cereals, chips, corn muffins, and taco shells—and sent them to a lab to get their DNA fingerprints.

The results came back a few weeks later. One of the packets, of Kraft Foods' taco shells, tested positive for StarLink DNA. Bohlen's group announced their discovery at a press conference on September 18, and the agbiotech industry was suddenly at the center of another emergency. Although StarLink represented less than 1 percent of the U.S. corn harvest in 2000, the StarLink discovery led the *CBS Evening News,* and the next day stories of tainted tacos and contaminated tortillas appeared in papers across the nation.

There was no one to blame but the EPA. They had approved StarLink for feed only, at a time when any Midwestern farmer could have told them it was impossible to segregate grains. Some grains were bound to mingle, either in the mechanical harvesters, or in the trucks that took the grain to the granaries, or in the granaries, or in the containers that shipped the grain on the final leg of its journey to the processing plant.

Over the next six months, seventeen people complained of having allergic reactions after eating taco shells. One woman went into anaphylactic shock after eating three enchiladas made with corn tortillas. The cases were investigated by the Centers for Disease Control in Atlanta and the Food and Drug Administration, which said they could find no evidence that StarLink protein was responsible. But critics charged that seventeen people complaining of allergic reactions was too small a sample to sound the all clear.

In the meantime Aventis had voluntarily withdrawn the product and started to pay out millions in compensation to farmers. The USDA bought back hundreds of thousands of bags of corn seed that contained traces of Cry9c, at a cost of $15–20 million, to maintain a stable market. Japan and Korea halted imports of StarLink for animal feed. In a new evaluation, the EPA cautiously decided that Cry9c had a "medium likelihood" to be a human allergen. The combination of the level of the protein and the amount of corn found to be commingled posed a "low probability" of sensitizing individuals to Cry9c.

On March 9, 2001, the EPA announced that it would no longer split registration for human and animal food. In the future, whatever corn was fit for Daisy to chew on had to be fit for humans as well. The tainted taco scare was over. The industry could return to the monarch butterfly.

By September 2001, drafts of the six papers that the EPA had asked for on Bt corn and the monarch were ready for distribution to the media. The scientific reports concluded that there was no immediate significant risk to the monarch from the two most commonly grown Bt corn types, Bt11 and Monsanto's Mon810, supporting the earlier results rushed through in 1999. Moreover, the studies showed that monarch caterpillars would have to be exposed to pollen levels on milkweed leaves greater than 1,000 grains per square centimeter before they would show any toxic effects. And although caterpillars were indeed found on milkweed during the one or two weeks when pollen is shed by corn, pollen levels on milkweed leaves were found to average only about 170 pollen grains per square centimeter in cornfields.

Reports from several of the studies showed much lower concentrations, even within the cornfield. Overall, the researchers estimated that "fewer than one percent of all North American monarchs would be affected by doses of [Bt] pollen high enough and at the right time to even see a subtle growth effect." [29] The *New York Times* headline declared, "Data on Genetically Modified Corn Reports Say Threat to Monarch Butterflies Is 'Negligible.' "

However, Knockout Bt corn—the one insiders called Event 176—was judged to be harmful to monarch larvae at concentrations of only 10 grains per square centimeter. Novartis, the producers of Knockout, had never managed more than a 2 percent market share; the company announced that Knockout would be phased out by 2003.

To groups such as the Union of Concerned Scientists, the research on Event 176 was a disturbing example of how the EPA had not been doing its job properly. Jane Rissler of the UCS said the monarchs had "lucked out" because Event 176 was not popular with farmers.[30] The industry had also been lucky that their estimates for the toxicity of Bt corn to the monarchs had essentially been borne out.

Losey and Obryki were not about to withdraw, however. In their joint paper for the EPA, they reported that the monarch larvae might still be in danger from eating pollen mixed with the corn plant's anthers—the male organs that produce the pollen. The anthers, it turned out, were much more toxic than the pollen grains. The scientists urged the EPA to grant only a one-year extension for Bt corn—until further research could throw more light on the anther issue.

Another researcher, Mark Sears of the University of Guelph in Canada, disagreed. His study found only whole anther parts on milkweed leaves and these, he said, would be too big for the caterpillar to eat. "To a caterpillar an anther is about as big as a city bus," he said. "Maybe some of the larger caterpillars eat them, but we haven't seen any evidence of that."[31] That debate would continue.

Chip Taylor of Monarch Watch continued to worry about the long-term effects of the Bt pollen. He warned that the monarchs might survive but be harmed in some way—possibly by suffering weakened digestive systems. They might be unable to fly the long migration route or perhaps be unable to reproduce in the spring. In the end, however, Taylor conceded that Bt corn was "probably not" the monarch's greatest hazard; it was more likely the weather. Monarch populations take a severe dip during droughts. From information gathered partly by the national monarch fan club, he had estimated that the monarch population had fluctuated from twenty-eight million wintering in Mexico in 2000 to nearly one hundred million a year later.[32]

In October 2001, the EPA reapproved the five Bt corn types on the market for five more years. Losey and company had lost a battle, but at the same time the EPA asked for more research on long-term effects. Green groups kept up the fight, complaining that the EPA was at fault for not having paid more attention to the plight of the monarch before approving Bt corn. The EPA replied that regulating pesticides is always a hazardous business; as one EPA official put it bluntly, "You can't test everything."[33]

The monarch was also facing hazards other than Bt corn and drought. Janet Anderson, director of the EPA's Biopesticides and Pollution Prevention Division, concluded that chemical pesticides "are killing monarchs at a far higher rate than Bt corn pollen is." Margaret Mellon of the Union of Concerned Scientists was still worried about the corn anthers; she thought the EPA should have delayed its approval until the anther research was complete.

But most of those involved—academics, industry, and other environmentalists—thought the "process" of the industry getting together with academics had been valuable—"a blueprint for how to do research in the public interest," as one of the researchers put it.[34]

Dr. Mellon agreed. "This was a model way to go about getting information on whether or not a risk exists. It brought scientists, environmental and government folks together with industry, found a pot of money, set a research agenda, got proposals, funded the research, and got it done before [EPA made] a decision about renewal. This was a really important process that should be followed routinely by the government as it makes decisions about GM products—and it's not."

THE PLANT HUNTERS

The greatest service which can be rendered to any country is to add a useful plant to its culture.

—THOMAS JEFFERSON

The Russian botanist, geneticist, geographer, explorer, and linguist Nikolai Ivanovich Vavilov was also the greatest plant hunter of all time. Between the two world wars, when Russia was in revolutionary turmoil, Vavilov scraped together Soviet funds to launch more than two hundred expeditions to Asia, the Middle East, Africa, and the Americas to gather hundreds of thousands of botanical species. From the mountains, forests, and open fields of these distant lands, he sent back to his Leningrad plant institute rare varieties of staple foods—rice, wheat, corn, barley, oats, and potatoes—as well as lesser-known lentils, chickpeas, soybeans, a host of vegetables, nuts, fruits, and spices.

Sometimes he selected what he wanted from cultivated crops or open markets. At other times, as he wrote in a 1920s memoir of his expedition to Ethiopia (then Abyssinia), Vavilov had more adventurous forays. He described how he shot his way past "fifteen-foot crocodiles with gaping jaws" as he forded the Blue Nile in search of a peculiar variety of wheat and rare strains of barley.[1] Later, camping on the river's shore, he awoke to find the floor of his tent a seething mass of venomous black spiders and scorpions. The only way to get rid of them, he decided, was to lead them outside by the light of an oil lantern, a slow but eventually successful method. Still further into the African bush, Vavilov encountered armed bandits whom he immobilized in a favorite Russian fashion: he got them hopelessly drunk on five-star brandy and made his escape while they were sleeping it off.

Some of the plants he found were virtually unheard-of in the developed world, such as Ethiopian *tef*, a kind of millet used in making a spongy pancake called *injera;* a peculiar oil-producing plant with black seeds called *ramtil* or *noog;* and a wheat with violet grains. In Taiwan he collected medicinal plants more easily from rows of local healers' stalls in the market. In all, Vavilov accumulated the largest collection of food plant seeds in the world—an international gene bank that surpassed the collections of the famous nineteenth-century botanical gardens. None of Vavilov's rival plant hunters—British, French, German, Dutch, or American—could keep up with the energetic and single-minded Russian who dedicated his life to finding plants that could help to increase the world's food supply.

His exploits brought him scholarly recognition abroad and popular fame at home. "Vavilov Crosses the Andes," declared an *Izvestia* headline. "Vavilov Visits Japanese Scientists," reported *Pravda*. In 1926 the publication of his book, *The Geographic Origin of Cultivated Plants,* brought him the Lenin Prize, the top Soviet decoration for science.[2]

If Vavilov was the greatest plant hunter, he was also among the few to work his craft as a mission, a public service he believed would benefit all humankind. While Vavilov's immediate aim was to improve the backward and chaotic state of Russian agriculture by collecting seeds that would grow in the Soviet climatic zones, he had a much wider vision: to expand the food supply for all the world's population.

Vavilov's dedication to pure science—a devotion that ultimately cost him his life—stands in stark contrast to the modern Northern bioprospectors, often referred to as biopirates, most of whom serve only commercial masters. They work in a new world where publicly funded agriculture has been in steady decline; where farmers buy new seeds each year instead of saving them from their harvest, as they have done for centuries; where U.S. patent laws have allowed the protection of living organisms, including plants; and where plant breeders from the North now have an opportunity to make fortunes hunting new varieties in the biologically diverse regions of the South. Some of the more rapacious expeditions of these new plant hunters—and the monopoly patents they acquired on plant varieties scavenged in Southern lands—have so offended agricultural leaders in those developing countries that there have been urgent calls for a reform of the property rights system.

Vavilov's working life was a continuous celebration of a most important botanical insight—as significant in practical terms for agriculture as Mendel's discovery of genetics was important in theory. For the first time, Vavilov pinpointed the original locations where staple plants had been domesticated and, therefore, where a plant hunter could expect to find the centers of greatest diversity. The discovery of those "centers of origin," as he would call them, brought order and purpose to plant hunting for the first time. Before Vavilov, plant breeders had no idea that the greatest genetic richness of any given crop could be pinpointed in one place.[3] The question for Vavilov when he set out on this quest was where to look.

Archaeologists suggested that these hordes of botanical riches would be in the cradles of civilization, in the fertile valleys and crescents of large rivers such as the Nile, the Tigris and Euphrates, the Indus and the Ganges, and the Yangtse-kiang and Huang-ho. But when Vavilov went to those well-prospected sites, he found no extraordinary variety in the plant life. He reasoned that at the dawn of agriculture, early farmers might have tried to grow their food plants on higher ground, in places where they would be protected from wild animals and other humans who might be still hunting and foraging. In addition, mountainous areas have complex environmental conditions that can change with each hill or dip, resulting in a startling variety of plants within a very compact area.[4]

To develop his concept, Vavilov took his expeditions deep into highlands where traveling was immeasurably more difficult and hazardous. As he traveled the globe, Vavilov confirmed his theory, recognizing seven mountainous "centers of origin" of the major food plants—a number that other scientists would later expand to twelve. By 1940, Vavilov's seed collection at his Institute for Plant Industry in Leningrad housed more than 250,000 specimens. The collection was more than a seed bank. Vavilov tested the new varieties at hundreds of experimental stations throughout the Soviet Union—a botanical archipelago that became the envy of plant breeders everywhere.

Hitler, when he invaded Russia in 1941, created a special seed commando unit of the SS, the *Russland-Sammelcommando,* with orders to bring back Vavilov's collection.[5] The Germans returned empty-handed, of course, after failing to capture Leningrad, and incredibly, the collection survived the Nazi bombardment. The seeds were housed in the old czarist Department of Agriculture in St. Isaac's Square, which was left intact because it also contained the splendid Astoria Hotel, where Hitler had planned to hold his victory celebration. During the siege, Vavilov's workers had to defend his collection against the starving citizens of Leningrad who came looking for potatoes or even seeds to sustain them. Several of the workers died of starvation at their desks rather than touch the precious seeds. Such dedication could only have been inspired by a truly charismatic leader.

Scientists who encountered Vavilov on his travels spoke not only of his capacity for hard work but of his disarming personality, a contrast to the Western stereotype of a Russian. Always dressed in a suit, with a tie and a hat, Vavilov cut a dashing figure as he ventured into the fields to fill his satchels with interesting specimens. He was accompanied by a stenographer, a photographer, and often a mule caravan for his local guides and supplies or, where possible, a convoy of early automobiles. Given the difficulties of getting around and the chaos of the early days of the Bolshevik Revolution, the number of countries he visited is truly astonishing. But the young plant hunter—he was thirty when the czar was overthrown in 1917—was so single-minded in his botanical quest that he paid little attention to the political and social upheaval around him. Vavilov's researchers apparently adored their purist professor. To one of them, he explained his commitment. "I really believe deeply in science; it is my life and the purpose of my life. I do not hesitate to give my life even for the smallest bit of science."[6]

But Vavilov's freedom to practice his pure science did not last. Stalin needed an excuse for the failure of his collectivization of Soviet agriculture; Vavilov and his institute colleagues were easy scapegoats. Stalin found an alternative "Bolshevized" voice for Soviet agriculture in the son of a peasant farmer, Trofim Lysenko. By denying the existence of genes and instead offering a "socialist" method of creating improved strains by growing plants under special environmental conditions, Lysenko became Stalin's chosen agronomist. Vavilov, who had traveled extensively outside Russia and had met and worked with the early geneticists, knew that Lysenko was wrong. He refused to bow to Stalin's unscientific whim.

Even when the purge of his Leningrad plant institute began in the mid-thirties, Vavilov was unyielding. "You can bring me to the stake, you can burn me, but I will not renounce my convictions," he famously told a Lysenko follower.[7] One by one his colleagues were arrested and accused of being disciples of the new "bourgeois" science of genetics. For good measure, they were also charged with plotting to overthrow the revolution. Vavilov was among the last to be arrested—on a plant-collecting expedition to the Ukraine. He was tried as a traitor and sentenced to death, a ruling later commuted to twenty years in jail. The sentence would effectively be the same, however. Vavilov died of starvation in prison in 1943, aged fifty-five. Soviet agriculture would continue to suffer under Lysenko, and only after Stalin's death was Vavilov posthumously reinstated as a hero of Soviet science.

Today's bioprospectors or biopirates—depending on your point of view—continue to search for exotic plants in Vavilov's centers, either directly as corporate agents or indirectly through universities and agricultural institutes that make deals with agribusiness. Science and the altruistic yearning to feed the world are lesser motives for these modern plant hunters. Several of them have made the headlines, but for reasons other than scientific inquiry.

One such hunter was a Colorado bean merchant named Larry Proctor. Like Vavilov, he went searching in Mexico for rare plants—specifically in his case for new varieties of beans. But Proctor was pursuing a very different goal. He was not interested in seed banks. He wasn't concerned about collecting seeds for the common heritage of humankind or sharing his new varieties with other breeders or university researchers. To other bean merchants, he appeared to be bent on cornering the market on yellow beans in the United States. Antibiotech groups would brand him as a classic example of the new generation of biopirates—as audacious and outrageous as those who had tried to patent the neem tree, the ochre spice turmeric, and basmati rice from India. Others would say Proctor was only doing what the patent law allowed. Either way, Proctor and his yellow beans became very famous in the food world.

At the beginning of the 1990s, Proctor bought a bag of dry beans in a market in the northwest Mexican border state of Sonora. He took them home and picked out the yellow-colored beans, planted them, and allowed them to self-pollinate. With each successive generation, he says, he selected only the seeds that were yellow, until he had a "uniform and stable population"—just as Mendel had selected peas in his monastery garden.

By 1996 Proctor had bred what he would claim was a unique bean of the *Phaseolus vulgaris* botanical tribe, which includes kidney beans, pinto beans, and navy and black beans. What distinguished his bean from all the others, Proctor claimed, was that his bean had a distinctive shade of yellow, which he identified in the *Munsell Book of Color,* a standard reference work on colors. He named the bean Enola, his wife's middle name, and filed for a patent for his precious legume. At the same time he applied for a U.S. Plant Variety Protection Certificate. He said that the bean was most likely descended from the "*Azufrado*-type" varieties very popular with Mexicans in Sonora, where they have been grown for centuries.

Officially, patents are granted for inventions that are new, useful, and nonobvious. Plant variety protection certificates are awarded for varieties that are new, stable, uniform, and distinct. Proctor's Enola bean was granted both a patent and a certificate, enabling him to sue anyone in the United States who sold or grew a bean that he considered to be his particular shade of yellow. In fact, yellow beans of Mexican origin have been grown and consumed in the United States as far back as the 1930s. At the time, several U.S. companies were importing yellow beans from Mexico, and several Southern farmers were growing them.

In 1999 Proctor, exercising the rights of his new legal monopoly, sued two of the companies that were importing yellow Mexican beans. He demanded royalties of six cents a pound on all yellow beans entering the United States from Mexico. One of the companies was run by Rebecca Gilliland, who had grown up in Mexico eating the yellow beans she was now importing into the States, beans that had been grown in Mexico since the eighteenth century. Gilliland thought Proctor's suit was a joke. "How could he invent something that Mexicans have been growing for centuries?" she asked.[8] Gilliland blamed the patent office for granting Proctor a patent without enough evidence that he had grown something new. "Next time, Proctor should tell the patent office that he invented tortillas last night and I'm sure they would believe him," she said bitterly.

But Proctor was serious, and so the U.S. Customs Service had to act. Officials started holding up Gilliland's trucks of Mexican beans at the border to see if they were carrying any yellow ones. Her company began to lose customers in the United States, but a challenge to Proctor's patent would incur legal costs of at least two hundred thousand dollars. Gilliland had no choice but to stop importing the beans.

Mexican farmers were outraged that Proctor was trying to corner the U.S. yellow bean market and that they had lost money on their bean exports. For them, the Enola patent was also an affront to Mexico's cuisine. Beans are the principal source of vegetable protein for Mexicans; yellow beans have been eaten by the residents of Sonora for as long as anyone can remember. The varieties are even known by local names that suggest the color, like *sulfur* and *canario*.

Proctor's claim to have bred a new variety came under serious attack. The biopiracy watchdogs of the activist group RAFI discovered that there were scores of *Azufrado* bean varieties held at the publicly funded International Center for Tropical Agriculture (Centro Internacional de Agricultura Tropical, or CIAT) at Cali in Colombia, one of sixteen publicly funded international agricultural research centers.

Professor James Kelly, a bean breeder at Michigan State University and president of the Bean Improvement Co-operative, told RAFI that the Enola patent was "inappropriate, unjust, and not based on the evidence or facts." He suggested that Proctor may not have done enough to justify the granting of a patent. He had simply grown the beans, which are self-pollinating, and allowed them to reproduce themselves. "This is a routine procedure used by bean breeders to maintain the purity of stocks and varieties," said Professor Kelly.

In the patent, Proctor had claimed that he had created "a segregating population of plants," a requirement for a patent application. "This is incorrect," said Kelly. "He simply observed different plant and seed types. He planted a mixture of beans that were different in shape, size, and color. This is not a segregating population which must result from a cross-pollination. Simply growing and 'selfing' [self-propagating] a specific seed type hardly implies novelty or invention. All Proctor did was multiply something that already existed. It was not unique in any sense of the word. To patent a color is absolute heresy."[9]

In Colombia, CIAT challenged the patent as "both legally and morally wrong." Saying they had "solid evidence" that Andean and Mexican peasant farmers had developed the bean first, they filed a formal request with the U.S. Patent Office for a reexamination of the Enola patent.[10] CIAT's gene bank holds more than twenty-eight thousand samples of seeds of the genus *Phaseolus,* including at least twenty-five that were listed as *Azufrado* from Mexico. Almost all CIAT's seeds are designated "in trust" materials. Under the terms of the 1994 agreement between the sixteen international agricultural research centers, of which CIAT is one, and the UN Food and Agriculture Organization, "in trust" seeds are maintained in the public domain and are not allowed to be included in any intellectual property claim. CIAT's official request for a reexamination of the patent challenged all of Proctor's fifteen claims as invalid. In particular, CIAT argued that the apparent reliance on color for novelty would make "a mockery of the patent system."

In face of his critics, Proctor was defiant. His lawyer said, "There's a lot of talk about Mr. Proctor doing nothing, but he devoted five years to coming up with what is basically a new bean."[11] In November 2001, Proctor filed a lawsuit against sixteen small bean seed companies and farmers in Colorado, claiming that they were also violating his Plant Variety Protection Certificate by illegally growing his yellow beans. At the same time Proctor amended the original patent with forty-three new claims. A year later Proctor settled out of court with the Colorado companies. The terms of the settlement were not disclosed. The challenge to the patent continued.

The Proctor case rang alarm bells in developing countries—just as the neem tree and turmeric episodes had done. In several of the countries, agricultural leaders feared that the number of invasive patent claims like Proctor's could only increase as Northern seed corporations, or Northern pharmaceutical companies, became ever bolder scouting for loot in Vavilov's centers of diversity, or as he also called them, "the bastions of the fortress of the plant kingdom." The rising demand of biotech agriculture for rare and exotic genes would simply make matters worse.

In developing as in developed nations, discussions about genetic engineering had to this point focused on the risks to human health and the environment. None of the developing countries in Asia had given its farmers official permission to plant any significant GM food or feed crops. Bt cotton has been released for commercial use in India and Indonesia. China is actively pursuing its own Bt crops. In Africa, only South Africa has approved the commercial growing of any GM crops (Bt cotton and Bt maize). In the remainder of Africa, only Kenya has supported field tests of GM crops.[12] In South America, Argentina was quick to go ahead with several GM crops, while Brazil and Chile officially did not allow them. But developing countries were also deeply concerned about trade implications of the new technology. Consumer rejection of GM foods in Europe, for example, meant that food-exporting countries in Asia and Africa were keen to remain GM-free. They had seen what the GM boycott had done to American grain exports. Argentina imposed an effective freeze on new GM food and feed crop approvals in an effort to avoid losses in export sales to Europe.[13] Ultimately, developing countries feared that biotech companies might be able to engineer crops that grow in Northern climates to produce essentially tropical foods, thus threatening their traditional exports.[14] This was already a trend. The agricultural economies of these countries depended to a large extent on producing special commodities, such as lauric acid oils from palm trees, used in soaps and detergents and once found only in the tropics.

Thirty percent of the population of the Philippines is dependent for a livelihood on palm oil.[15] Canola plants, which grow in North

America and Europe, have been genetically engineered to produce lauric acid, a clear threat to the Philippine palm oil industry. Coffee was another developing world commodity under siege, at least in theory. The center of origin of coffee is the highland region of Ethiopia, but in North America, caffeine genes have been inserted into soybeans. As the Harvard evolutionary biologist Richard Lewontin asked provocatively, "Why not Nescafé from Minnesota?" [16]

There is little wonder that the voices of concern over biotechnology from the continents of Asia and Africa were vehement. Countries in these places face the problem of finding enough food for their expanding populations over the next fifty years. The Green Revolution doubled wheat yields in India, and the Chinese boosted rice harvest by two-thirds—probably saving more than a billion people from starvation. [17] Although the proportion of the world population that is chronically undernourished has been more than halved, there are still about eight hundred million people in developing countries living at near-starvation level. [18] "Food security" in countries where the bulk of economic activity is still based on agriculture is not only a matter of growing enough. The problem is also finding cash to pay outsiders for the shortfall.

In the rich and well-fed nations, the decision about whether to grow GM crops could be taken at leisure, but at the start of the 21st century developing countries were being invited to decide more urgently whether the technology should be a part of their future—and they were not ready. In Kenya, for example, the ecologist Hans Herren worried about the real meaning behind the biotech industry ad that declared with frightening arrogance, "Biotechnology is one of tomorrow's tools in our hands. Slowing its acceptance is a luxury our hungry world cannot afford."

To Herren, the ad meant that too much emphasis was being put on GM seeds and crops "to the detriment of more conventional and proven technologies that have been very successful and whose potential lies mostly unused in the developing countries." Herren worried that "the trend towards a quasi-monopolization of funding in agricultural development into a narrow set of technologies is dangerous and irresponsible. . . . It is only too obvious to concerned scientists, farmers, and citizens alike that we are about to repeat, step by step, the mistakes of the insecticide era, before it is behind us."[19]

In India, so far the country hardest hit by the "biopirates," antibiotech activist Vandana Shiva declared, "Nothing less than an overhaul of Western-style [property right] systems with their intrinsic weaknesses will stop the epidemic of biopiracy. And if biopiracy is not stopped, the everyday survival of ordinary Indians will be threatened, as over time our indigenous knowledge and resources will be used to make patented commodities for global trade."[20]

The groundwork for such a challenge was laid at the United Nations Conference on Environment and Development at Rio de Janeiro in 1992. In the thaw after the Cold War, countries turned their attention to several long-neglected environmental issues, such as climate change, the depletion of the ozone layer, soil erosion, water supply depletion, and the transport of hazardous substances. The aim was to restrict activities that threatened the earth—oil drilling, mining projects, deforestation, and new dams. But the hottest topic that year was the United Nations Convention on Biological Diversity, or CBD, a legally binding commitment for all those who signed up to stop "biopiracy" and secure the conservation and sustainable use of biological diversity. (In 2002, 183 countries, including the United States, had signed the CBD but the United States had not ratified the treaty.)

The CBD marked an important break with the past in three respects. Instead of regarding biological and genetic resources as the common heritage of mankind, the new treaty sought to protect local communities, mostly the developing countries that generate and are dependent on biological diversity. The CBD would give states sovereign rights over such resources. Second, it required signatories to protect and support the rights of communities, farmers, and indigenous peoples over their traditional varieties and plant knowledge. Third, it required the benefits from the commercial use of those resources to be equitably shared, thus curtailing unfair exploitation by the wealthier industrialized nations. Plant hunters would have to obtain official permission to gather samples, not only from the state government but also from the local community.[21]

At the same time, the pharmaceutical and seed companies opened international negotiations to protect their own access to these resources and to preserve their property rights to anything they discovered that they could turn into a more useful and marketable product. The CBD came into force in 1993, and two years later the World Trade Organization (WTO) was established to administer a global trading system. A coalition of corporations, including Monsanto, argued that property rights contributed to the promotion of technological innovation and should form part of the new treaty. The WTO agreement contained a property rights section known as TRIPS, for Trade-Related Aspects of Intellectual Property Rights. The TRIPS agreement said that biological resources, including microorganisms and microbiological processes, should be subject to intellectual property rights and that countries should also set up some form of protection for plant varieties. The aim of the international seed industry was to "harmonize" property rights among developing nations. Before TRIPS, for example, only two countries in Africa—South Africa and Zimbabwe—had functioning plant variety protection systems.

Developing nations and environmental groups widely condemned TRIPS as just another way of sanctioning the piracy of biological resources.[22] They also pointed out that the treaty clashed with the spirit and the letter of the CBD. The two main objections to TRIPS were against the patenting of life forms and the loss of farmers' and community rights. As required by the WTO, however, several developing nations began considering their own version of the TRIPS section calling for a patent system or its equivalent for plant varieties. A dozen nations representing 70 percent of the world's plant biodiversity—China, Brazil, India, Indonesia, Costa Rica, Colombia, Ecuador, Kenya, Peru, Venezuela, and South Africa—formed a front pressing for more equal trade rules on patenting.[23] Concerted opposition to TRIPS came from Africa, where the Organization of African Unity drafted proposals that were much less exclusive than current property rights. In their draft, patents on life forms and on biological processes would not be recognized. Farmers would be allowed to save part of the harvest for seeding the following year, and if a plant were protected, it could still be used as a genetic resource for research purposes at breeding stations. At the time of writing, the TRIPS section was still being negotiated.

In these treaty negotiations, the most urgent challenge facing the international community is in Africa, the continent that stands to gain—and also possibly to lose—more than most from genetic engineering. Since 1970, cereal yields in Africa have increased at only about half the rate of those in Latin America. UN projections show that while chronic malnutrition is expected to decline in Asia and Latin America, it will continue to rise in Africa.[24] According to those estimates, Africa may have forty-nine million undernourished children by 2020, a rise of almost 50 percent from the year 2000.

Africa is fertile ground for the heated debate about whether transgenic crops could bring the continent relief from hunger. Poor soil, high temperature, low rainfall, and pests are permanent issues. African climates also vary so considerably that it is a challenge to breed varieties that will grow from region to region. A map of the major agro-ecologies of Kenya, for example, shows six different climatic areas, each requiring a different variety of corn.[25]

In addition Africa has to cope with political instability, civil war, and a working population dying of AIDS. Roads are less than adequate; arable land is dwindling; poverty is widespread. African farmers desperately need to increase yields. Africa's crop production per unit area of land is the lowest in the world. The production of sweet potato, a staple crop, is less than half the global average. A single acre of farmland in Europe produces with added chemicals six times the cereal grain harvested from an acre in Africa. Pests and disease account for 30 percent of African yield losses.

During colonial times, European powers established plantations to feed home markets, mostly with cash crops such as coffee. In postcolonial Africa, Western technology has not offered much help to the African farmer. The Green Revolution hardly touched Africa because the modern varieties were too costly and improved crop varieties, produced for temperate zone agriculture, were not suitable for Africa's soil, pests, and precipitation patterns.

Until recently the big seed companies had little interest in Africa. South Africa and Zimbabwe is a $300-million seed market, but the rest of the sub-Saharan seed market is so far worth only $200 million. Industry analysts say that introduction of GM crops could increase the entire African market by 50 percent.[26] But the seed companies will not operate in countries where there is no strong protection for property rights.

Africans fear a repeat of what happened during the Green Revolution, when seed companies developed high-yielding varieties requiring costly inputs for those who could afford them, leaving the majority of small farmers without improved varieties. Or they worry that the seed companies will continue to operate only in South Africa and Zimbabwe, the export-oriented horticultural markets of Kenya, and the emerging fruit markets of Egypt and Morocco, rather than the less lucrative markets.[27]

Cuts in public research funds have meant that African agricultural research stations already established in such countries as Kenya and South Africa have had to look elsewhere for support. One option has been to go to the biotech industry, a solution that tends, in the view of the antibiotech forces, to come up with projects that will help the embattled industry rather than the majority of local farmers. The result is two polarized models of African agricultural policy—one that includes transgenics and another that favors a more conventional approach using traditional breeding, management of soil fertility, and crop protection.[28]

The African debate has raised powerful voices on both sides. The Kenyan biologist Florence Wambugu is her nation's chief probiotech campaigner, as impressive on the stump as the Indian activist Vandana Shiva is for the other side. For the last decade Wambugu has worked on a transgenic sweet potato that has a built-in resistance to the feathery mottle virus that can reduce yields by up to 80 percent. The fact that her project was a collaboration between the Kenyan Agricultural Research Institute, USAID, and Monsanto has laid her open to charges of being a Trojan horse for industry.

"Some people say I am fighting for the company," Wambugu says in response to groups like Greenpeace. "But I believe the technology has benefits for people. . . . I'm not saying that transgenics alone will solve all the problems. But it will lead to millions more tons of grain. . . . In Africa GM food could literally weed out poverty. . . . In Africa most weeding is done by women, [so] reducing that would have a major impact."[29] Wambugu complains that aid workers have been "brainwashed" by Greenpeace and other antibiotech civil society groups into believing that transgenic crops are unsafe and will ruin traditional varieties in Africa. Moreover, she believes that Europeans are being force-fed "half-truths" about the dangers of biotech crops.[30]

One of Africa's strongest antibiotech voices is an Ethiopian, Tewolde Berhan Gebre Egziabher, who became head of his country's Environment Protection Authority in 1995. Passionately against allowing patents on living organisms, Tewolde has been involved in drafting the Organization of African Unity's response to the TRIPS property rights section. Like many Africans, Tewolde is worried about losing rare clusters of biodiversity; Ethiopia is the Vavilovian center for barley as well as coffee and tef. But more generally Tewolde does not trust the seed companies, or governments of the North, to look after the needs of the poor. There are no profits in Africa. "It's not the nature of genetic engineering that's the problem, it is the way genetic engineering has evolved," he has said. "Early on it came under the control of the private sector and is now being developed almost entirely by the private sector. By definition, the private sector's goal is to make money. It will not focus its attention on the needs of the poor, except as a way to sell its products."[31]

When Nikolai Vavilov first reached the Ethiopian highlands in 1927, he was especially taken by the small millet called tef and the spongy, slightly sour pancake it made. The plant, prolific and more nutritious than wheat, is the most important food crop in Ethiopia. Vavilov was not the first European to find tef. In the mid-1880s, British botanists had taken tef seeds back to the Royal Botanical Gardens in London. From there the plant was distributed to the colonies—India, Australia, and South Africa. Tef remains a peculiarly Ethiopian plant, but there is a U.S. Plant Variety Protection Certificate on a tef variety.

The certificate is owned by Wayne Carlson, an American biologist. He worked for the Ethiopian government in the 1970s, and when he left he took some seeds home to Idaho. He now grows tef on two hundred acres in a harsh dry valley on the Idaho-Oregon border and sells his grain to the Ethiopian population of the United States. Whether tef has a great future as a food outside Africa, other than among expatriate Ethiopians, is hard to tell. It might become attractive because, being essentially gluten-free, it could serve as a bread substitute for people who are allergic to wheat flour.

Carlson says he has no plans to use his tef plant certificate to challenge the Ethiopians if there should suddenly be an international tef fad. When Carlson started growing the grain, he used to send some back to Ethiopia for trials. One year, during a drought, he donated more than twenty thousand pounds to a relief agency. If Vavilov had survived Stalin's purges, it is the kind of gesture he himself might have made from his Leningrad seed bank, sharing the expertise and wealth of his botanical treasure which, after all the years of upheaval, is still intact in St. Isaac's Square.

THE CORNFIELDS OF OAXACA

A good aggressive bunch of American agronomists and plant breeders could ruin the native resources [of Mexico] for good and all by pushing their American commercial stocks.
—CARL SAUER, UNIVERSITY OF CALIFORNIA, 1941

The hope of the industry is that over time the market is so flooded [with genetically modified organisms] that there's nothing you can do about it. You just sort of surrender.
—DON WESTFALL, A BIOTECH INDUSTRY CONSULTANT, 2001

Zacateca Indian farmers in the high sierras of southern Mexico still work the land where their ancestors raised the distant relatives of America's dining favorite, corn on the cob. These rocky fields surrounded by cedar forests are far away from the high-tech laboratories of Europe and the United States, yet in the fall of 2001 the Mexican hilltop farmers suddenly found themselves on the front line of the international biotech wars. As the farmers harvested their latest corn crop, two researchers from the University of California at Berkeley reported that the genetic purity of the treasured native criollo corn had been contaminated by alien genes from transgenic varieties grown in the United States. The Mexican government's environment ministry declared the contamination to be "the world's worst case of contamination" of traditional farmer varieties by genetically modified crops.[1]

Environmental groups labeled the new research evidence of the nightmare of genetic engineering—the loss of ancient gene pools that provide breeders with genetic insurance against plagues and pests. Greenpeace declared that the alien genes in the criollos had not only ruined a potential source of irreplaceable genetic material, but had brought on an invasion that went beyond just agriculture. The presence of foreign genes was an affront to Mexican culture and sovereignty, as insulting as "if they had torn down the cathedral of Oaxaca to build a McDonald's over it."[2] Media headlines talked of "contaminating" and "tainting" the local varieties, as though the transgenic corn carried a new and possibly fatal infectious disease.

In their report in the scientific journal *Nature,* the Berkeley researchers, Ignacio Chapela and David Quist, had described two separate events. The first was the cross-pollination of an unidentified transgenic corn with a local criollo (a variety cultivated by local farmers without interbreeding with modern varieties, known generically as a landrace). The second event was more complicated and raised the possibility of far greater consequences. The two researchers reported that the genes from the genetically modified pollen were now unstable in the genome of the criollo, implying that these wandering genes might produce all manner of unexpected and destructive results.

Yet even as Chapela and Quist published the discovery, their university colleagues denounced the second part as a misinterpretation of the results. The two researchers were accused of being scientifically incompetent and ideologically motivated. One critic charged that their report had "more mysticism than science."[3] They were accused of raising an alarm without cause—like Dr. Pusztai with his snowdrop-laced potatoes and Dr. Losey with his poisoned monarch butterflies.

In a stunning reversal, *Nature* bowed to the critics and disavowed the legitimacy of the research, announcing that there had not been "sufficient evidence to justify" publication of the report. It was the first time in the 133-year-history of *Nature* that the London-based journal had withdrawn support for an article in defiance of its authors and their referees.

The industry was greatly relieved; another biodisaster seemed to have been averted. Biotech industry flaks branded Quist and Chapela fanatics in the antibiotech cause. "We believe that *Nature* erred in publishing the article to begin with, and it seems they came to the same unavoidable conclusion," said Val Giddings of the Biotechnology Industry Organization. "The authors made mistakes that first-year grad students learn to avoid, which further demonstrates that their commitment was not to data and science but to a religious commitment to an [antibiotechnology] dogma."[4]

What might have become another botanical debate about how to preserve landraces turned into an intense and at times vicious academic row. The Berkeley researchers complained of being "bullied" by the editors of *Nature* and "intimidated" by their colleagues, who in turn they accused of acting as agents of the biotech industry. On the Berkeley campus, there was talk of "neo-McCarthyian [sic] tendencies."[5]

In Mexico the matter became the subject of a national debate between those who were basically supportive of biotech—mostly from the agriculture ministry—and those who were skeptical, from the environment ministry. One of the pioneers of plant biotech, Luis Herrera Estrella, added his own dimension to the battle. Now director of Mexico's leading center for plant biotechnology, Herrera Estrella had worked on the first genetic transformations using *Agrobacterium tumefaciens*, the bacterium in common use—and employed by Potrykus and Beyer in golden rice—for inserting foreign genes into plants. But instead of following his colleagues into the then high-flying biotech start-up companies, Herrera Estrella had returned to his native Mexico to work in the government-funded agricultural research institution. He argued that the presence of one or two transgenes in the criollo corn would be unlikely to cause the disappearance of the native varieties. "There is no evidence that this represents a threat for the maize biodiversity of Mexico," he said. "For decades, the creole varieties have lived together with commercial varieties, including the hybrid varieties from the multinational companies, without causing their disappearance and in most cases not even their substitution by small farmers."[6]

Officially the contamination of transgenic maize should not have happened. For the previous three years, in an effort to protect their ancient varieties, the Mexican government had forbidden the planting of GM corn. If the Berkeley researchers were correct, farmers had either knowingly defied the ban on planting GM crops, or the modern corn genes had somehow slipped over the border in undocumented alien seeds and into the Mexican gene pool.

Trying to make sense of the academic bickering, reporters traveled to the remote hills an hour and a half's drive from the city of Oaxaca where the Berkeley team had taken their corn samples. They quickly discovered one possible explanation for what had happened.[7] Poor farmers in the village of Calpulalpan had been buying imported corn at the local government store, but they did more than eat it. Some farmers had planted the seed, hoping that it would produce a better yield than their local criollo varieties. In fact Mexican corn experts had known for years—long before the 1998 ban on GM plantings—that poor farmers all over Mexico had been sowing superior corn seeds from the United States brought back by migrants working in the American South. A second possible source of modern U.S. varieties was the corn imported from the United States into Mexico to make tortillas for the U.S. market. If any of this corn found its way to Mexican farmers they could not know whether the tortilla corn kernels had been genetically modified.

Sixty years after the Green Revolution, 80 percent of all corn grown in Mexico still consisted of local criollos, planted year after year from a farmer's harvested seed. The heady days of the Rockefeller program in the 1940s and '50s, when Mexico became self-sufficient in corn, were long gone. Six million tons of corn were imported from the United States each year—from grain elevators that did not separate GM from non-GM types. This was commingled grain that U.S. farmers could not sell to their big markets in Europe and Japan because of bans on GM products. The grain, including modified varieties, was being sold to Mexico and elsewhere and being sent to Africa as food aid.

The Oaxaca farmers, who were only trying to improve their existence by planting a few kernels from the store, were caught in a controversy created by an American grain industry that was unable to separate GM from non-GM grains. U.S. grain exports could not be guaranteed free of genetically modified grain. The Mexican government denied importing GM corn, and officials may have convinced themselves that they were getting GM-free shipments, but how would they know? The StarLink affair had demonstrated that no U.S. grain shipment is pure. As a Canadian biotech industry consultant, Don Westfall, warned about the spread of GM seed, "The hope of the industry is that over time the market is so flooded [with genetically modified organisms] that there's nothing you can do about it."

In the village of Calpulalpan, subsistence farmers were confused and worried by news of the Berkeley research results. They had been caught defying the ban against planting GM corn and feared that the government might burn their fields or even prosecute them. When they heard the word "contamination," they didn't understand its implications. They began to worry that some kind of poison was in the corn that they had planted in the hope of improving their harvest. Now they wondered if the foreign corn could harm their chickens or even their families. One farmer said, "I feel guilty. But another woman told me she planted it too. We don't know the damage we can do." [8]

What the public badly needed was an open debate on the merits of the Berkeley research and its implications for the future of one of the world's staple crops. If what Chapela and Quist had found was shown to be correct, there were serious questions about gene flow from genetic engineering. In the cornfields of Oaxaca, more was at stake than academic reputations. But *Nature* left the battlefield, telling readers to "judge the science for themselves."

Dr. Ignacio Chapela, a forty-two-year-old Mexican microbial ecologist, had not started out to find contaminated corn. He was a fungus specialist; his initial interest in the southern Mexican state of Oaxaca had been to set up a small genetic testing station for local farmers. Chapela had helped the farmers identify and test local wild matsutake mushrooms. The project had been successful; local villagers began supplementing their meager incomes by selling the mushrooms to the Japanese.[9]

At the same test station, Chapela's research student, David Quist, set up a project to show villagers how to test for possible gene flow from transgenic varieties in order to determine whether the government ban was working. Local farmers had apparently been planting the "new" corn for several years.[10] They said that the seeds, which arrived in government trucks for sale at state-subsidized community stores, were so robust that they would grow anywhere, even through cracks in the sidewalks. Diconsa, the national subsidized food program, distributed corn to twenty-three thousand stores all over Mexico. The local farmers said that the imported corn kernels were larger, had a lighter color, and were not as sweet as the local kernels. They yielded three times as much grain as the traditional varieties, at least initially. Over time, however, the imported varieties often proved susceptible to local plagues.

Quist bought kernels from the local government food store, assuming that the corn had come from the United States and would probably contain alien genes. The store corn tested positive for GM, as Quist had predicted, but so did the local criollo. Quist took samples of both types of corn back to Berkeley for a more sophisticated analysis.

In the laboratory, the store sample showed what appeared to Quist and Chapela as a strong presence of the genes used in the creation of pest-resistant Bt corn—including the controversial 35S promoter gene from the cauliflower mosaic virus. Fifteen of twenty-two criollo samples would also test positive for traces of transgenes. In the *Nature* report, Quist and Chapela concluded that there had been "a high level of gene flow from industrially produced maize" to the local land-races.[11] From their sample, the two researchers estimated that between 1 and 10 percent of the native corn in the Oaxaca fields might contain transgenes. However, since the samples were from remote areas, the researchers speculated that higher rates of gene flow could be expected in regions that had larger, more accessible farms. They did not doubt that Mexican farmers had been planting transgenic corn in defiance of the government ban. The question was, how long had they been doing so? Long enough, perhaps, for the transgenes to become permanent fixtures in the criollo genome and be passed from one generation to the next.

Quist and Chapela also reported seeing fragments of genes from the 35S promoter at unexpected sites in the criollo's genome. This suggested to them that unstable snippets of foreign genes were moving about inside the corn genome—the conclusion that would set off the firestorm at Berkeley. The idea of transgenes "jumping around the genome," as one of the Berkeley critics put it, was one that "would have changed some of the basic assumptions of biotechnology."[12] Such a suggestion was also highly political because it supported the contention that bioengineering was an inherently imprecise and risky technology.

Chapela and Quist fully expected some opposition from the industry and also from their colleagues. In 1998 Berkeley's Department of Plant and Microbial Biology had signed a five-year deal with Novartis, the Swiss biotech company (now merged with the British Zeneca to become Syngenta). The deal gave Novartis the right to pick the best plant research at the university's department in exchange for a $25-million grant over five years. The agreement was controversial with several academics, including Quist and Chapela, who felt that the department had simply sold out to corporate interests. There were campus demonstrations against the Novartis compact.

In defense of the arrangement, the university argued that with public funds for research drying up, a financial link with industry was the only way Berkeley could compete with other research institutions in the new biotech world. But as a result of the deal, the Berkeley plant biology department became a target for antibiotech activists. On the night of October 11, 2000, activists destroyed GM maize being grown by department researchers. Even though Quist was in Mexico at the time, he came under suspicion among his colleagues as the staff member who might have told the activists which plot contained the GM corn. He was quickly exonerated from this charge by the university, but the suspicion lingered on, according to Quist. When he and Chapela published their work, some critics linked Quist and his partner to these earlier disagreements.

In the wider scientific community, few actually doubted that Quist and Chapela could have found transgenic DNA in criollo corn. Maize is an open-pollinating species, which means that maize plants readily exchange pollen with other maize plants growing nearby, a characteristic recognized long ago by local farmers as a means to adapt varieties to their own preferences and ecology. As a result, of course, Mexican criollo has always been a constantly evolving species. It is certainly not the same plant that was growing, say, a hundred years ago. Without new genes the criollo becomes inbred over generations and loses vigor, just like other corn varieties and other plants. The Mexican farmers describe that when a variety "gets tired" *(se cansa)*, they deliberately seek out other maize varieties to improve the gene pool.[13] In short, genetic diversity in a farmer's field, even high up in the sierras, is not a static condition.

Because no one doubted that Mexican migrant workers returned home with improved U.S. corn varieties to mix with their local criollos, Quist and Chapela's colleagues made light of the discovery of contamination. One Berkeley researcher, Nick Kaplinsky, said that the chances of some transgenic corn being planted in Mexico—despite the government ban—is "kind of an obvious no-brainer." He noted that in India farmers had planted transgenic cotton illegally, and in Brazil they had planted transgenic soybeans, although Brazil has also officially banned GM crops. "Farmers will plant the best seed they can get, especially if they're subsistence farmers," he said. "If you're surviving on a harvest and you have a choice between corn that will give you one kilo [of grain] per plant or two kilos per plant, you'll take the two-kilo-per-plant seed, presumably."[14] Kaplinsky suggested that soon someone would "come up with good scientific evidence that [transgenic corn] is growing all over [Mexico]."[15]

On the more combustible conclusion—that fragments of transgenic DNA were jumping around in the genome—the Berkeley critics dismissed Quist and Chapela's results as bad science. In a letter to *Nature*, researchers from the Berkeley plant department charged that Quist and Chapela's claim was "unfounded" because they had wrongly interpreted their analysis, probably by looking at false positives from lab contamination.[16] Another letter, from a group of researchers at the University of Washington including Matthew Metz, formerly of Berkeley, said, "The discovery of transgenes fragmenting and promiscuously scattering throughout genomes would be unprecedented and is not supported by [the] data."[17]

In their reply, Quist and Chapela acknowledged that they had misidentified two of the DNA sequences but claimed that their paper reporting the gene flow was still unique.[18] Under less trying circumstances, as the British weekly *New Scientist* pointed out, a partial retraction of the paper might have been enough to satisfy both sides. But after hearing from the critics, *Nature* demanded that Quist and Chapela retract the whole paper. They refused. At that point *Nature* took the extraordinary step of disavowing their work. *Nature*'s editor, Philip Campbell, insisted that his journal had never said that Quist and Chapela's conclusions were wrong. "We have said that they are not convincing on the basis of the evidence that we have published." Campbell denied that a campaign against the two researchers influenced his decision to disavow the paper.

Chapela would not back away from the second conclusion, about the uncontrolled movement of transgenic DNA. "To those who would like to bury [this] reality," Chapela said, "I can only echo the words of Galileo, *'Eppur si muove.'* " Galileo is said to have muttered these words ("And yet it moves") after renouncing to the Inquisition his theory that the earth revolves around the sun.[19] Quist and Chapela continued to believe the industry had masterminded a campaign to discredit their work—"an assault on the very foundation of science," as Chapela put it.[20] Whether the assault was on the foundation of science would take a while to unravel. There was certainly an instant campaign against Quist and Chapela.

On the day their paper was published, Internet probiotech forums carried immediate demands for the paper to be retracted. One of these forums, AgBioView, with an e-mail list of thirty-seven hundred scientists, led the attack. The forum is moderated by Professor Channapatna Prakash, a professor of plant molecular genetics at Tuskegee University, Alabama, and an outspoken biotech advocate. One correspondent claimed the paper was "junk science that shouldn't have made it past rudimentary peer review process."[21] An early criticism came from a "Mary Murphy" who attacked Chapela for being on the board of directors of Pesticide Action Network, a group trying to reduce the use of pesticides, and so, claimed Murphy, "not exactly what you'd call an unbiased writer."

Murphy's posting was followed by a message from "Andura

Smetacek." Smetacek had appeared on AgBioView before in a rant about Losey's experiments with monarch butterflies and how green groups were using a PR firm to create scare campaigns about transgenic crops. Smetacek claimed, incorrectly, that Quist and Chapela's paper had not been peer-reviewed. She said Chapela was "first and foremost an activist." A British antibiotech activist traced Murphy and Smetacek's electronic personas to the Bivings Group, a Washington, D.C., public relations company that had Monsanto as a client, but Bivings denied any knowledge of either name.[22]

The hint of industry character assassins lurking in cyberspace was all some green groups needed to convince them that Quist and Chapela had indeed been victims of a long-standing industry campaign against GM "dissidents"—including Arpad Pusztai and John Losey. To these groups, any researcher who found evidence that questioned the efficacy or safety of bioengineering would be hounded by agribusiness. The view of the biotech industry was that first Pusztai, then Losey, and now Quist and Chapela were all alarmists whose work should never have been published without further inquiry.

The result was a serious breakdown in scientific discourse on this vital issue. Scientists are expected to disagree. Research papers that pass through a peer review are sometimes found to have errors in them after publication. Such papers are supposed to generate inquiry that eventually settles the issue. But the highly charged debate over biotechnology was having a distinctly different result. Enormous efforts were being expended on propaganda by both sides in the war. The green groups, aided and in some cases abetted by the media, were promoting research results about the possible hazards of biotech well beyond their worth. In return, the companies were devoting huge resources to discrediting such research. But the same effort was not going into examining the questions raised. Biotech crop breeders argued that such studies were, generally speaking, a waste of time, especially in the face of increasingly scarce resources.[23]

Public funds for basic agricultural biotech research had all but dried up, as Berkeley had recognized when the school accepted funds from Novartis. Less than 1 percent of the U.S. Department of Agriculture's biotech program—$1.6 out of $250 million—was committed to risk assessment and the safety testing of such devices as the 35S promoter, the toxicity of snowdrop lectin, or the possible extinction of the monarch butterfly from eating Bt corn. (The allocation would be doubled in 2003 after a battle between Ohio Congressman Dennis Kucinich and the biotech industry.)

The constant wonder was that these potential hazards had not been addressed before the release of GM crops into the environment. But from the start the FDA and the EPA had passed the research buck to the companies. The regulatory policy for GM foods created under President Ronald Reagan and continued under presidents Bush and Clinton was to let the companies do the research: if they found anything wrong, they would tell us. The policy had ultimately caused a steady decline in the public trust in biotech science.

In science journals the anonymous peer-review process is supposed to distinguish good science from bad, letting only the best into the public record to be used as reference for future inquiry. Peer reviews are not always perfect, of course, but *Nature*'s disavowal of its original publication of Quist and Chapela's work threw that professional code into turmoil. "The specter of unseen actors manipulating events is especially worrying," commented the British weekly *New Scientist*.[24]

Some scientists complained that *Nature* had left its readers in the lurch by not reporting the private rows behind the widespread condemnation of Quist and Chapela's work. Following convention, *Nature* had not identified the referees who had supported the original paper, nor the one subsequent referee who had apparently persuaded *Nature* to disown the article after publication. Similar questions had been asked—and left unanswered—about the peer reviewers of the Puzstai paper published in *The Lancet* and about the Royal Society reviewers who had concluded that Pusztai's work was flawed. But the question was, where should such inquiries stop? For example, should

Nature, as some scientists suggested, have revealed how much of its advertising revenue comes from biotech companies in order to prove its independence on GM issues?

In the biotech wars, the scientific journals found themselves in an increasingly uncomfortable position, buffeted by a strident public reaction in Europe and North America unmatched in the recent history of technology. As Philip Campbell, the hapless editor of *Nature,* wrote, "It must have been Murphy's law that ensured that our technical oversight, embarrassing in itself, was in relation to a paper about one of the most hotly debated technologies of our time."

This "technical oversight" had allowed Quist and Chapela to become instant martyrs in the biotech wars for a brief, confusing, and in the end for them humiliating moment. If the aim of the two Berkeley researchers had been to raise public alarm beyond what Chapela called "some anecdotal evidence"[25] about the risks of gene flow from transgenic crops, they had certainly succeeded. Their peers had dismissed the science of their experiment and had made light of their "discovery" of the "contamination" of native corn varieties, but that did not prevent their conclusions from being written into antibiotech lore. Their contribution reignited the key environmental debate over transgenic plants that had been glossed over by the agbiotech companies in their rush to get the new products to market.

Long before the advent of biotechnology, scientists had been discussing what might happen when genes flowed from modern cultivated crops either to landraces nurtured over generations by peasant farmers or to the plants' wild relatives. Researchers expected two potentially harmful consequences.[26] First, the new transgenes might confer new fitness or defenses to the wild plants, creating a kind of superweed. Second, the new genes might cause the extinction of the landrace or wild species, which might otherwise provide new, invigorating genes to keep the species alive.

In 1992 Calgene scientists in California warned, "The sexual transfer of genes to weedy species to create a more persistent weed is probably the greatest environmental risk of planting a new variety of crop species."[27] The greatest concern was about centers of diversity for staple crops. What would happen when pollen from GM varieties pollinated corn landraces or wild species in Mexico? The rice bowls of Asia? Or the original potato fields of Peru? The talk was never *if,* only *when* this contamination would occur.[28]

In the summer of 1995—five years before Quist and Chapela started their research—the Mexican government and the International Maize and Wheat Improvement Center (CIMMYT) held a seminar on gene flow to which they invited experts from the United States who had been studying Mexican corn. One of the experts, Major Goodman of North Carolina State University, was asked about the risk that the "rural poor" in Mexico would plant new varieties. He had no hesitation in saying that, of course, they would plant any variety that helped them improve their yield. "I don't doubt in any way that all sorts of remote Mexican farmers are going to grow transgenic crops, and I think they are going to do it whether it is legal or not." He added, "The same thing is probably true all over the world. You have everything from Mexican migrant labor to Mexican Ph.D. students to missionaries in the Congo. All of these people think that they are doing good by carrying this material around. . . . Somehow . . . the problem has to be faced worldwide."

Another corn expert, Garrison Wilkes, a professor of biology at the University of Massachusetts, who had spent thirty-five years researching the spiky relative of maize known as teosinte, warned of the difficulty of preserving the ancient gene pools. "I can confidently say that, with few exceptions, these populations will not exist thirty-five years from now." The only hope was to end what he called "habitat displacement" from changes in land use. "The single most significant effect on pushing teosinte to extinction is barbed wire. That is, more intense land use through grazing."

Since plants can only be fertilized by pollen from the same species, GM radishes can mate with non-GM radishes but not with non-GM carrots. The problem is that nearly all the major crops, such as corn, rice, barley, and sorghum, have close relatives that are regarded as weeds somewhere in the world and could theoretically be turned into superweeds.

Weeds reduce the growth and yield of crops by competing for water, light, and nutrients or by producing toxic compounds and even harboring insect pests and plant pathogens. The term *weed* is subjective, of course. No plant is born a weed; it becomes one only if it happens to be in the wrong place at the wrong time—as far as humans are concerned. Lawn grass, that staple of suburban life, is a weed in a farmer's field. The Weed Science Society of America defines a weed as "any plant that is objectionable or interferes with the activities or welfare of [humans]." Weeds are also expensive. Farmers and gardeners spend billions of dollars a year trying to control weeds.

A big problem for crop farmers arises when a wild relative mimics the crop. Wild varieties could adapt by gene flow so that they begin to look and act like the cultivated plant. Although they are still really weeds, they are difficult to distinguish from the crop. Examples of such mimics are found among rice, sorghum, corn, and millet. This phenomenon is generally not a problem for staple crops in Canada and the United States—where native foods include only a few types of berries, sunflowers, the Jerusalem artichoke, pecans, black walnuts, and the muscadine grape. However, minor crops, such as radishes, can potentially have problems. The wild radish is a common California weed, while the cultivated radish is an important California crop.[29] Radish pollen floats easily on the wind and also rides across the fields courtesy of California's native insects. One study showed that any weed within a meter of cultivated radish plants was doused with pollen; a low level of pollination was detected at one kilometer.

Gene flow could be a particularly serious problem in centers of crop diversity, such as India and Mexico. In India, rice breeders troubled by wild rice mimics developed purple varieties of cultivated rice to differentiate them from the wild rice. But the wild rice still won the battle. The flow of genes from the crop to the wild rice produced a new variety of wild rice that was as purple as its cultivated cousins.

In some cases the transformation could work in reverse. A cultivated variety might adopt the traits of a wild species. In Europe, gene transfer from wild sugar beet to its crop relative was a serious problem for Europe's sugar beet harvest. A dominant gene for *bolting*—sudden flowering of the plant—flowed from the wild relative and caused the crop beets to flower in their first year, making them unusable for commercial sugar production.[30]

Studies also show how gene flow can result in the extinction of a rare or wild species. The hybrid—a cross between a rare or wild plant and a cultivated variety—can become "depressed" and lose its vigor in the next generation (as was seen in hybrid corn). It can become so weak that it dies out.

Another route to extinction is known as *gene swamping,* which occurs when the genome of a rare or wild species is "swamped" by genes from a cultivated relative that repeatedly pollinates the wild plant season after season.[31] Studies show that such repeated mating between a common, cultivated species and a rare one can send the rare species into extinction. Either through loss of hybrid vigor or through swamping, a plant can disappear in three generations.[32] Gene flow from cultivated cousins to wild ones has been implicated in the extinction of at least six wild crops—including relatives of hemp, corn, pepper, and the uplifting little sweet pea.[33]

Corn, one of the most promiscuous of the staple crops, is an obvious candidate for gene flow. Corn pollen can travel surprising distances on a good wind or with an insect. Government regulations suggest that farmers need to leave a buffer zone two hundred meters wide to avoid cross-pollination.[34] Corn experts could easily forecast that the criollo was bound to become mixed over the years with genes from improved varieties from the United States.

Teosinte, the wild variety of corn found in Mexico, is believed by most experts to be the ancestor of domesticated corn. Teosinte contains many genes that are potentially useful in corn breeding, but the kernels of teosinte itself are inedible. In human terms, it's a weed. Ecologists worry that corn pollen from a transgenic crop might fertilize a teosinte plant and pass on genes that would enable teosinte to mimic domesticated corn and thus become an agricultural nuisance.

In some Mexican fields, teosinte has already developed the red plant color, hairy leaf sheaths, and wide leaves typical of cultivated corn, and often escapes weeding.[35] The wild variety looks so much like the improved variety that it's often hard for the subsistence farmer on a small plot to distinguish between the two. Certainly, a mechanical harvester on a larger farm would not be able to tell the difference.

If the biotech industry had been reluctant to discuss gene flow before the Quist and Chapela affair, the subject was now a matter of open and international debate. Environmentalists argued that sooner rather than later, the world's crop landraces were all at risk of being contaminated by transgenes unless the centers of diversity of the major crops were turned into special reserves—national agricultural parks, quarantined from transgenic plants. The probiotech lobby countered that such gene flow was both inevitable and probably even beneficial.[36] Farmers who had cultivated landraces for centuries were constantly improving their crops by planting seeds of new varieties next to old ones and then selecting the desired offspring.

In Mexico, where maize is the staple food, particularly for the rural poor, small-scale farmers strive to enhance their landraces by planting modern improved varieties alongside their traditional crop.[37] Dr. Major Goodman, the corn geneticist at North Carolina State University, believes that any new gene is likely to shuffle through strains of corn with no harm done. "If it's detrimental, it will be eliminated rather quickly. If it's beneficial, it will stick around and multiply a bit, and it might lend a little bit of protection to populations that are currently rather endangered, as a number of these populations are."[38] As Matthew Metz, the University of Washington biotech researcher, put it, "Farmers should not be relegated to the role of museum keepers of static 'traditional varieties.' Numerous international seed banks keep stores of important crop diversity."

The question is whether the seed banks themselves can stay pure. Mexico has the most important depository of maize genes in the world at the International Maize and Wheat Improvement Center (CIMMYT). Initial tests on corn seeds going back to 1967 found no traces of telltale 35S promoters. However, as the guardian of corn gene diversity, the center came under attack from environmentalists who thought that Chapela and Quist's results were evidence that the center had somehow failed as the custodian of the Mexican maize gene pool.

In their own defense, the center's researchers pointed out that even if the Berkeley research was correct and transgenes had passed into local criollos, that did not necessarily mean that the landrace gene pool had been permanently depleted. Mendel's genetic theory suggested that the transfer of a single gene carrying a trait for insect or herbicide resistance—if, indeed that is what had occurred in Oaxaca—should on its own have little impact on genetic diversity.

For example, if a modern American yellow maize variety, such as those imported as food grains into Mexico, carried a Bt transgene for pesticide resistance and was planted in a field with local traditional white corn, the two types would mate, exchange genes, and after a while, perhaps a few generations, the following plants would emerge: There would be plants with yellow grains and the Bt gene, plants with white grains and the Bt gene, plants with yellow grains and no Bt gene, and some with white grains and no Bt gene.

So although there had been gene flow from a new GM corn to an ancient farmer-nurtured variety, the center's researchers argued that the original variety had not, in fact, been lost. Meanwhile, overall genetic diversity had actually increased. Whether this was a good way of increasing diversity, or a viable way of maintaining criollos, depended to a large extent on the farmer. If the yellow corn varieties were producing better yields than the traditional white varieties, then the farmer might reasonably discard the white types altogether—following the same agricultural practice used by farmers through the ages. The problem was that no long-term research existed on the relative significance of the two key factors—gene flow and farmer selection.

Finally, studies showed that the impact of gene flow on wild ancestors such as teosinte, the closest relative of maize, might also be limited. Modern corn genes can certainly flow via pollen into teosinte, but studies showed that they do not necessarily swamp the teosinte genome, suggesting that some kind of genetic barriers may be at work preventing such a takeover.[39]

The heated debate about landraces led inevitably to the complex legal question of patent rights. When alien genes were discovered in the Oaxaca criollos, the Mexican farmers had been fearful of government sanctions, but the government ban was not the only potential legal violation. The biotech seed companies were gradually extending their biotech patents into developing countries. If the Oaxaca hill farmers were found growing transgenic corn in violation of these patents, they could, in theory, be sued by the U.S. seed companies for royalties. It might seem absurd for a big multinational like Monsanto to sue a Mexican hill farmer, but the company had already demonstrated its determination to litigate to protect its intellectual property.

Monsanto had already either threatened or taken legal action against hundreds of farmers in Canada and the United States whom they accused of using their proprietary seeds without permission. The case of Percy Schmeiser, a canola farmer of Saskatchewan, Canada, was a warning to any farmer who thought of sneaking some genetically modified seeds past the watchful eye of Monsanto's seed police.

Past retirement age but still farming, Percy Schmeiser had lived an uncomplicated life on the Canadian prairie. For forty years he had grown crops of bright yellow rapeseed on his fourteen-hundred-acre family farm. He had been an assemblyman for the province of Saskatchewan and a mayor of his local township; in his spare time Schmeiser had climbed Mt. Kilimanjaro and tried three ascents of Everest. In 1998 he was looking forward to giving up the farm, the politics, and the mountain climbing to go fishing. Then suddenly this rugged son of the Canadian Midwest found himself accused by Monsanto of being a seed pirate. Monsanto charged him with illegal planting of the company's genetically modified, patented canola—the new name for rapeseed used to make cooking oil. The company's canola plant had an alien gene that made it resistant to Monsanto's popular herbicide Roundup.

Unknown to Schmeiser, private investigators working for Monsanto's seed police had taken seed samples from his 1998 canola crop, had them analyzed, and reported to the company's St. Louis headquarters that more than 90 percent of Schmeiser's crop consisted of Monsanto's Roundup Ready canola seeds. In the company's book, Schmeiser looked suspiciously like another seed thief.

Any farmer growing Monsanto's canola, or its Bt corn, was required to sign a "technology-use" agreement and pay the company fifteen dollars an acre. In 1996, the first year the GM canola seeds went on the market, six hundred Canadian farmers signed up; four years later the number had grown to twenty thousand. These farmers produced nearly 40 percent of the canola grown in Canada.[40] With the aim of recovering their estimated $250-million research and development costs, Monsanto vigorously enforced the contracts, even encouraging farmers to snitch on neighbors who were flouting the rules.

The company provided a toll-free telephone number for farmers willing to turn in their neighbors. They were asked to "Dial 1-800-ROUNDUP and tell the rep that you want to report some potential seed violations or other information. It is important to use 'land lines' rather than cellular phones due to the number of people who can scan cellular calls. You may call the information in anonymously, but please leave your name and number if possible for any needed follow-up."[41]

The company received hundreds of tips. Offending farmers received letters from Monsanto's lawyers, threatening legal action with the option of a confidential out-of-court settlement. Percy Schmeiser refused to play the company's game.

Schmeiser did not deny that his 1998 crop contained plants that had grown from Monsanto's super-canola seeds, but he claimed that the company's seeds had trespassed on his land by one of three possible routes. The conventional seed he bought from the store might have been contaminated. Seeds could have fallen off a passing truck carrying a neighbor's GM canola on the way to market and landed on his farm. Or genetically modified canola pollen from a neighboring farmer's land could have been borne on the wind, or carried by bees, and pollinated his conventional crop. In a combative move, Schmeiser countersued Monsanto for $4.2 million. He charged that the company's private detectives had trespassed on his land and that Monsanto had contaminated his crop and defamed him. Schmeiser accused the company of "arrogant, high-handed, and shocking conduct and callous disregard for the environment." Far from being a seed thief, he maintained, he was a victim of the new technology invading his property.

The case was closely watched by seed companies and environmentalists alike. All farmers and plant breeders know that rapeseed, or canola, is a promiscuous plant of the *Brassica* group. Left to its own devices, canola would "spread pollen all over the countryside," [42] according to one canola expert. In court Schmeiser admitted that at least some of his crop consisted of Monsanto's GM plants. However, his estimate of the portion of his canola plants that were genetically modified was more like 60 percent, not the 90 percent Monsanto claimed. For its part, the company argued that "forces of nature such as wind and bees are clearly insufficient to produce a 90 percent crop of Roundup Ready canola." [43]

Monsanto wanted to make an example of Schmeiser. The company spared no expense in aggressively prosecuting the case. They hired an expert in road vehicle aerodynamics to see if canola seeds really could fly off a passing grain truck onto Schmeiser's land. The expert, having developed a theoretical model using local weather conditions and prevailing winds, estimated that the maximum distance a canola seed could fly would be 8.8 meters from the road. There was a stretch of public land 16.8 meters wide between the road and Schmeiser's field, suggesting that the canola seed could not have come off a passing truck. The Monsanto expert conceded, however, that Schmeiser's crops dipped into that public land area; it was possible that the Monsanto samples had been taken from some plants that were within the pollen flight distance. Schmeiser's lawyer suggested that small prairie whirlwinds known as dust devils might have carried the pollen further, but Monsanto countered that such phenomena are about as rare as lightning strikes.[44]

The judge took nine months to reach the conclusion that windborne pollen or seeds from a passing truck were almost certainly not the source of Schmeiser's Roundup Ready canola plants. But the source really didn't matter, ruled the judge. Whether Schmeiser's Roundup Ready seeds had come as whole seeds from the store or from a neighbor, or had fallen off a truck, or had been created in his conventional canola by pollen flown in by bees, or had blown in on the wind was not the issue. The fact that Schmeiser "knew or ought to have known" that he was growing Monsanto's patented canola was an infringement of that patent and put the responsibility squarely on him to pay the company a royalty.

Monsanto and other seed companies celebrated the ruling as a reassuring sign that they would be able to continue to extract a "technology-use" fee for their products, but farmers and environmentalists were stunned. Where would the liability end? they asked. "This means that farmers with land close to genetically modified crops will have to pay royalties to the companies for products they never purchased and got no benefits from," commented Margaret Mellon of the Union of Concerned Scientists.

Schmeiser was ordered to pay Monsanto $19,000 in damages and more than $150,000 for the company's legal fees. Instead Schmeiser appealed. The Canadian federal appeals court upheld the lower court's three key rulings. First, the mere presence of the patented plants on Schmeiser's land was a patent infringement. Second, a farmer who knows or ought to know that the Roundup-resistant variety is on his land and who saves the seed and reuses it, as Schmeiser had done, has infringed the patent. Third, even if the farmer makes no money from the GM seeds—and Schmeiser had not made any extra money from his contaminated crop—he is still acting illegally. Schmeiser's lawyer argued throughout the case that Schmeiser had bred his own brand of canola over forty years and that this ruling took away the rights of farmers to replant their own seed. In the fall of 2002 Schmeiser decided to take the case all the way to the Canadian Supreme Court, where a decision was not expected for another year.

Schmeiser's battle with Monsanto turned him into a folk hero of the antibiotech forces. Green groups paid for him to travel the world to tell his compelling David and Goliath story. In India he was presented with the Mahatma Gandhi award, given in recognition of nonviolent work done for the betterment of mankind. And his case brought into focus several wider legal issues to do with seed contamination. In the future, big seed companies like Monsanto could find themselves in court as defendants, not prosecutors.

For example, organic farmers whose crops are contaminated by GM pollen stand to lose their organic certification—a prized possession that takes several years to acquire. And farmers of conventional crops whose plants are accidentally contaminated by GM seeds or pollen from the back of a passing truck or a fierce gust of wind may no longer be able to sell their product as GM-free and therefore lose that market. In such cases the legal question is whether seed companies can be held liable for "polluting" farmers' fields with genetically engineered organisms.

Experts acknowledged that the promiscuous *Brassica* genus, which includes canola, cabbages, cauliflower, and radishes, should be given a wider berth than self-pollinating crops such as rice and wheat. Also, some pollen grains stray a lot farther than others, riding more easily on the wind because they are smaller or lighter.[45] Some grains are fertile for a whole day or so, others only for a matter of minutes. Bees carrying pollen can fly over ditches and hedges, as can birds carrying seeds. Animals can roam long distances after eating a genetically modified crop; having not fully digested the seeds, they can then deposit them even farther away than either the wind or bees could carry them.

Lawyers wondered whether the manufacturer of the seed could be considered as the owner or person in control of the "pollutant."[46] Saskatchewan organic farmers filed suit against Monsanto and Aventis for contaminating their canola crops in Western Canada.

A potentially more serious legal liability, for farmers or companies, is presented by "third generation" transgenic plants that will "grow" pharmaceuticals. Researchers are working on corn plants that will produce cancer-fighting antibodies, edible vaccines, and human proteins for therapeutic purposes. Norman Ellstrand, professor of genetics at the University of California, Riverside, and a leading expert on corn genetics, worries about gene flow from such plants. "They will pose special problems if we do not want those chemicals appearing in the human food supply," he has warned.[47] Farmers on the Canadian prairie and in the American Midwest wonder who will be blamed if a vaccine for smallpox ends up in a packet of breakfast cornflakes.

In the end, the problems of gene flow highlighted by Quist and Chapela in a Mexican criollo, and claimed by Percy Schmeiser in his Canadian canola, stem from the same sources—the arrogance of corporate control and the failure of government regulations. Biotech companies such as Monsanto sold genetically modified canola, knowing that their seeds could contaminate the crops of Canadian farmers growing conventional, unmodified canola plants. U.S. grain handlers and food processors sent corn for tortillas and other foods to genetic hot zones like Mexico's Oaxaca cornfields with full knowledge that some of the corn was likely to be whisked out to nearby fields and planted in violation of Mexico's ban on GM seeds.

The U.S. government failed to limit the sale of such corn to Mexico. The Mexican government imported the corn for the nation's subsidized food program surely knowing that the American grain was bound to be contaminated with GM varieties. The Zacateca Indian farmers, Percy Schmeiser, and for that matter the Berkeley researchers had no say in the invasion of these genetic modifications from the American agricultural establishment.

So Shall We Reap

One thing is sure: the earth is more cultivated and developed now than ever before; there is more farming but fewer forests, swamps are drying up and cities are springing up on an unprecedented scale. We have become a burden to our planet. Resources are becoming scarce and soon nature will no longer be able to satisfy our needs
—Quintus Septimus Tertullianus, 200 b.c.

Across southern Africa in 2002 the harvests failed, leaving almost 15 million people facing starvation. Drought one month, floods the next destroyed crops across the continent, but AIDS and local political turmoil had also exacted their toll. As boreholes went dry and crops withered, the world saw the all-too-familiar pictures of women and children lining up for their daily handful of grain or flour offered by nations with plenty. The African famine of 2002 had a new dimension, however. Three countries—Zimbabwe, Mozambique, and Zambia—made the astonishing decision to refuse food from the United States containing genetically modified seeds. Fear of the new seeds was so great that leaders—presumably well-fed leaders—decided to chance their luck and look elsewhere for help.

How did the fear of GM foods rise to this tragic level? A decade after Americans had eaten their first GM food—a harmless tomato that ripened more slowly—farmers had planted genetically modified seed on more than 130 million acres worldwide. But biotech companies had failed to convince the international community outside the United States—even nations in Africa on the brink of starvation—that these novel crops were safe for humans and the environment.

Under pressure from UN relief agencies, Zimbabwe and Mozambique agreed to take the U.S. corn, providing it was milled and free of seeds that might be planted. The Zambian government stubbornly refused the aid altogether. "Simply because my people are hungry, is not a justification to give them poison," declared Zambia's president, Levy Mwanawasa.

From the American perspective, the Zambian decision looked churlish and irresponsible. A small, undeveloped nation of 10 million with traditional agricultural methods, poor soils, and an inhospitable climate was putting nearly 3 million people at risk of starvation by turning away food. Moreover, this food was being offered by the world's most powerful industrialized nation, a country with the most technologically advanced agricultural system and the most safety-conscious consumers.

The U.S. corn that might have gone to Zambia, or any other nation, unavoidably contained genetically modified grains. One-third of the corn planted in the United States is genetically modified, and because it is not separated in the American grain system, American food aid corn is not guaranteed to be GM-free. In 2001 the UN's World Food Program, which distributes aid donated by individual countries, fed these GM foods to 52 million people.[1] The United States offered unmodified wheat or rice, but the Zambians only wanted corn.

From an African perspective, however, the Zambian decision looked quite different. First, the Zambian government did not believe that it had made an irresponsible decision. There was no question of letting its citizens die of starvation. There was plenty of non-GM corn in the world's granaries and the Zambians believed they would be able to secure enough of it with donated aid funds from elsewhere. Second, Zambia was not against GM crops per se, but the government had been advised by its top scientists to favor the precautionary principle—which basically meant that until a food was proved safe, it was off-limits. The Zambian scientists had come to this conclusion after seeking advice from experts on both sides of the debate in Europe, the United States, and South Africa. The scientific group had concluded that GM food was still a potential health hazard and, citing the recent Mexican gene flow example, that American corn could contaminate local African varieties. Among the familiar health concerns cited by the Zambians were that GM foods could produce unpredictable toxins or new allergens and that antibiotic-resistant marker genes were still being used in America and could potentially cause harm. In addition, the Zambian scientists noted that while millions of Americans may consume corn in processed foods such as cornflakes and taco chips, Zambians do not eat corn as a staple food. In Zambia, unprocessed corn is the staple food and usually the only source of carbohydrate.

A third reason for rejecting the U.S. aid was that Zambia, like other African countries, exports agricultural products to Europe and European consumers were basically anti-GM. Until now, Zambia had remained GM-free and the government was concerned that if it allowed the U.S. corn into the country that their farmers would be tempted to plant the new seeds as well as eat them. The country's corn crop would then be contaminated in European eyes, and Zambian exports might suffer—even though the exports were mainly horticultural and did not include corn.

The fourth reason was that Zambia, again like most African nations, still lacked a system of internal regulations for monitoring and testing GM crops and products. For several years, the developing countries had sought a way of regulating the import and cultivation of GM crops and foods. In 2000 they had succeeded, against U.S. opposition, in securing an amendment known as the Biosafety Protocol to the 1993 Convention on Biological Diversity (CBD). This amendment gave governments the right to regulate GM foods.[2] The protocol required exporters of GM seeds for planting to give the importing

country written notification of their arrival. Yet, there was no such obligation for crops used in processed foods or for grains intended for direct human or animal consumption. Thus, the United States had no
obligation to notify the Zambians, or any other country, that they
were sending food aid that may contain GM corn. Indeed, the U.S.
had been sending such corn as food aid for several years.[3]

The incident quickly escalated into a full-blown diplomatic row.
America accused the cautious Europeans of persuading the Africans
that genetically modified foods might be unsafe. In turn the Europeans
suggested that the Americans were cynically trying to shove corn they
could not sell elsewhere down the throats of starving Africans, and calling it charity. EU officials went to Zambia to explain that if Zambia
grew GM corn, it would not affect the country's ability to export other
agricultural products—vegetables, flowers, and coffee. Those products
would be unaffected because they don't mate with corn.

The U.S. Secretary of Agriculture, Ann Veneman, blamed antibiotech forces for scaring Zambians into believing that GM corn
would harm them. "It is disgraceful that instead of helping hungry
people, these individuals and organizations are embarking on an irresponsible campaign to spread misinformation and create an atmo-
sphere of fear, which has led countries in dire need of food to turn
away safe, wholesome food."[4] While Veneman's target appeared to be
such staunch biotech opponents as Greenpeace and Friends of the
Earth, the United States was also threatening to declare a trade war on
Europe for its four-year moratorium on the approval of new GM
products. The ban was hurting American farmers; the American government was expected to appeal to the World Trade Organization,
charging protectionism.

Longtime antibiotech campaigners were quick to pick up on the
cynical view of the U.S. aid. Hope Shand, of the antibiotech ETC
Group (the Action Group on Erosion, Technology, and Concentration), formerly RAFI, said, "The U.S. and the biotechnology industry
have been desperate to show the benefits of this technology. Now they

are trying to sell the product by giving it away."[5] The science journal *Nature* picked up on another aspect of the U.S. donation. U.S. food aid grants and loans are only available for the procurement of grain from U.S. farmers. The journal noted sarcastically the "extent to which aid donors like to enjoy most of the fruits of their own benevolence"; American farmers would be receiving "a few dollars more on top of the billions being lavished on domestic farm support."[6]

In many ways the biotech industry and the U.S. government had only themselves to blame for this latest fiasco. Since the beginning, while the industry claimed that their products would save the world from malnutrition, seed companies created only crops that made money for themselves and the wealthier farmers who could afford the premiums. Even Western consumers were yet to receive a direct benefit from these novel foods.

Bound by its "substantially equivalent" doctrine—which declared the new foods safe because they were substantially equivalent to the old ones—the U.S. government had told consumers that a transgenic tomato was just like an ordinary tomato, even though bioengineers acknowledged that there were substantial differences. Ignoring the distinctions, the U.S. government and the grain merchants had not required farmers to separate their harvest into GM and non-GM grains, so when it came to offering starving nations food aid, there was essentially no choice. Zambia's decision polarized the debate, leaving both sides looking as though they had bungled the affair. As the Zambian agricultural minister, Guy Scott, told *Time* magazine, "I don't think there are any particular heroes or villains in this whole thing, it's just a balls-up."[7]

In the brief, turbulent history of biotech agriculture, the Zambian famine also turned into another bitter contest for public opinion. This time, however, millions of people were on the brink of starvation while the two sides engaged in yet another war of words. Once again, a quest for the scientific truth of GM foods was undermined by special interests. The debate became the most poignant in a long list of events that had eroded public confidence in the new crops—golden rice, the cornfields of Oaxaca, potatoes with snowdrop genes, the monarch butterfly, patents on basmati rice and yellow Mexican beans, StarLink corn, and the mysterious escape of canola genes on Percy Schmeiser's farm in Saskatchewan. The Zambian incident also refocused attention on the developing world as the new front line in the biotech wars.

In North America, the birthplace of biotech, the revolution was stalled by 2003. Farmers could not sell GM crops in several international markets and were reluctant to consider new biotech products. In America 35 percent of the corn crop and 75 percent of the soybean crop was GM, but worldwide, the figures dropped dramatically—to 36 percent of soybeans, and 7 percent of corn. The European market remained bleak. Consumer opposition was still high; the EU was about to introduce strict labeling rules for GM foods. The outcome of the British farm-scale trials of GM crops—the most comprehensive tests so far of the effects of these crops on the environment—was expected in the summer of 2003. Even governments, such as Japan's, that had allowed GM imports slowed their approval after the StarLink disaster—when GM corn approved only for animal feed was found mixed with corn for humans. And the United States and Canada were both postponing commercial planting of GM wheat because of market jitters. A Canadian study suggested that any big wheat exporter stood to lose a third of its wheat market if it started to plant GM wheat.[8]

In addition, there was still uncertainty over the supposed benefits from GM crops. Did Bt crops cut down on pesticide use? Did Roundup Ready crops reduce the overall use of herbicides? The answers depended on who did the measuring. According to an industry survey for 2001, transgenic crops have been a success. The report said herbicide-tolerant soybeans saved U.S. farmers $1 billion and a GM variety of corn raised yields by 1.58 million metric tons.[9]

But an independent researcher, Charles Benbrook, who has followed the use of these crops for the Northwest Science and Environmental Policy Center, challenged the industry figures, arguing that they represented only tiny savings to U.S. agriculture as a whole. Benbrook said the soybean farmers didn't spend $1 billion less by using Roundup Ready. The supposed saving of $1 billion represented the estimated extra cost to GM farmers of using alternative weed killers to Monsanto's Roundup. But, he argued, farmers who don't use Roundup find other, cheaper ways of controlling weeds, including tilling their fields. According to Benbrook, Roundup users probably only break even on GM soybeans.[10] The Bt corn figures were right, but one variety's gain represented less than 1 percent of the 250 million tons of corn grown each year.

Overall, it would appear the gains have been marginal. Roundup Ready crops have reduced the average number of active chemical ingredients applied per acre but have modestly increased the average use of actual herbicide. Bt corn has had little impact on pesticide use. In any case, Benbrook says, whether GM crops reduce pesticide use is the wrong question. The real question is whether biotech can be used in a more subtle way to strengthen plants' defense mechanisms and put an end to the "pesticide treadmill" that occurs when pesticides destroy beneficial insects and, at the same time, create new, resistant pests requiring ever more pesticides.

Several studies continue to show a risk that GM crops will interbreed with wild relatives and thereby not only create GM-tainted crops but also sprout "superweeds." A team of researchers at Ohio State University showed that wild sunflowers, considered a weed by many U.S. farmers, become hardier and produce 50 percent more seeds when crossed with a GM sunflower resistant to a particular moth larva.[11] Researchers in France found gene flow between GM sugar beets and wild cousins.

Some companies have deliberately avoided such dangerous liaisons by not producing transgenic products that are promiscuous and have wild relatives growing nearby. Monsanto has not tried to improve sunflowers, which are native to the United States, for example. Even proponents of genetic engineering have warned that certain crops, such as the randy canola, might not be suitable for all fields. Planting on Canada's prairies, where sufficient space could be found to create an effective refuge between GM and non-GM crops was fine, but canola could not be trusted in more confined environments, such as Britain.

In America, an entirely new gene flow risk emerged in *biopharming*. Pharmaceutical companies realized that they could make medically important proteins more cheaply in the kernel of a corn cob than in fermentation factories. Farmers immediately saw a new and potentially profitable niche market. But grocery manufacturers took fright at the very idea. They envisioned green groups finding a gene for diarrhea in a taco shell. "Who wants a pharmaceutical in their cornflakes?" asked Rebecca Goldburg of Environmental Defense.[12] The biotech industry promised not to biopharm in major corn-producing states such as Iowa, Illinois, Indiana, and Nebraska, but grocery manufacturers wanted stricter assurances, especially after the summer of 2002, when their fears were realized.

The USDA found GM corn containing a pharmaceutical protein growing in two soybean plots in Iowa and Nebraska. The offending corn had grown from seeds left over from the test crop planted the year before by a Texas-based company, Prodigene. Exactly what genes were found and what drugs they were for remained a company secret. The USDA ordered the burning of 155 acres of surrounding corn and the quarantine of half a million bushels of harvested soybeans from the test field. The accident was potentially more disastrous than StarLink—as the food processors made clear. If the GM corn had been in a cornfield, not a soybean crop, there could have been cross-pollination. Food processors and grocers, foreseeing the possible ripple effect of such scares, were fearful of losing international markets for their popular brands.

Meanwhile the biotech companies—including Monsanto, Syngenta, Bayer, and DuPont—were doing their best to overcome the disastrous launch of biotech agriculture. Monsanto's new president and chief executive, Hendrik Verfaillie, a chemist from Belgium who rose up through the company's ranks, promised to behave "honorably, ethically, and openly" in the future, in contrast to the arrogance admitted by his predecessor, Robert Shapiro. Monsanto offered its knowledge in the rice genome for public use. But the new image did not help the company's fortunes. Monsanto's agbiotech business was badly affected by Europe's moratorium and Brazil's rejection of biotech. The company, which was still the biggest crop biotechnology firm, saw its share price cut in half during 2002. In the first nine months of 2002, the company's sales plunged more than 18 percent, to $3.45 billion from $4.25 billion. At the end of the year, Verfaillie was forced to resign.

In addition, Monsanto's leading herbicide, Roundup, was reported to be losing the battle with some weeds that had evolved a resistance to it. The company's Roundup Ready corn and canola seeds, which were resistant to the herbicide, were the cornerstone of the company's food crop business. The new weeds were not "superweeds" in one biotech sense because they had not developed their resistance as a result of gene flow from a transgenic crop, but simply by evolution. But the lesson was clear—sameness can be a plague in agriculture whether it be mono-crops or mono-herbicides. Monsanto's rival, Syngenta, seized the opportunity to push its own products, suggesting that farmers should not limit themselves to one type of weed killer, as many had been doing with the successful Roundup.[13]

Independent biotech research continued, however. From the labs came word of several new products—from rice that maintained its yields when grown in cold, dry, or high-salt conditions that would kill normal plants, to tomatoes that acted as medicines, to potatoes that produced more protein.

The new rice was developed by a team at Cornell University.[14] The researchers experimented with a sugar named trehalose, found in a rugged desert plant known as the resurrection plant. During periods of drought the plant looks as though it has died, but after a rain shower it springs back to life. Its revival is attributed to the presence of trehalose, which is thought to protect plants in salty, arid, and cold conditions by maintaining the right balance of nutrients and minerals needed for photosynthesis. The Cornell researchers found that a pair of genes that made trehalose, borrowed from the common bacterium *E. coli*, produced the sugar in a variety of basmati rice. The growth rates of the basmati rice were just 20 percent below normal when the plants were exposed to salty, cold, or dry conditions.[15]

Meanwhile British and Dutch scientists were working on a GM tomato that produced flavonols, powerful antioxidants that fight disease by neutralizing harmful oxygen molecules that circulate in the body, damaging tissues and accelerating the aging process.[16] The researchers discovered a gene in the common petunia that produces the enzyme that makes the flavonol. The taste of the antiaging tomato was not affected, apparently.

Elsewhere there was still much apprehension about transgenic plants. Planting of GM crops was still illegal in many developing countries, not only for food safety reasons but also for international trade purposes; governments like Zambia wanted to preserve their official GM-free status.[17] When India decided to allow GM cotton to be grown in 2002, farmers saw more risk from international corporate control of seed markets than from harmful gene flow. They still believed, for example, that although Monsanto had renounced use of the Terminator technology, the company might still be able to make seeds sterile and thereby deprive Indian farmers of their traditional right to save and replant seeds from their own harvest.[18] The antibiotech lobby continued to be preoccupied by concern over this genetic trick. Although both Monsanto and Syngenta renounced the Terminator, the technology behind it was not abandoned.

And after the Terminator came the Exorcist. This was a method of killing off alien genes at the end of the plant's life cycle so that they do not appear in the pollen or the seeds and, therefore, cannot be passed to a wild relative or the next generation. The method was immediately dubbed "The Exorcist" by the masterful headline writers of the action group ETC. The technology uses a little enzyme that automatically snips off all the genes spliced into a plant at a particular stage in its development—for example, at an early stage of the development of the fruit before the pollen starts to ripen and become active.[19] The success of the method rests in the timing, of course, and some scientists are skeptical that it could ever be totally reliable. It might excise all the foreign DNA from the plant fruit but not from the seeds. The antibiotech forces saw the Exorcist more as Terminator II, a "greenwash of the issue, rather deceptive," was how the environmental group Greenwatch U.K. described it.[20] Certainly the Exorcist could be seen as just another way of preserving the seed company's intellectual property rights.

Weary from the biotech wars and with new and more detailed knowledge of plants' genomes, researchers began to take another look at traditional breeding. Tinkering with the plants' own genes, awakening slumbering genes already there rather than introducing new ones, was an attractive route that defused the "Frankenfood" argument. Because no alien genes were transferred, the new plant could not be labeled transgenic or GM.

The first plant genome to be sequenced was that of the small mustard plant, *Arabidopsis thaliana*, often used as a model for crop plants. It has prompted researchers to look at the genes that tell the plant exactly when to flower; the genes that govern plant height, root length, or the size of flowers, leaves, and seeds; and especially the genes that help the plant's natural resistance to hungry insects and creeping blight. Making plants flower earlier could extend the growing season for grains and fruits, perhaps enabling farmers to grow more than one crop a year. Even small advances in flowering time could help rice farmers. Rice needs just over six months to grow before it can be harvested, so speeding up the flowering time could allow two crops. Making plants flower later would stop vegetables such as spinach and lettuce from bolting too soon—sending up stems that sap energy.[21]

Other researchers have tried to influence a plant's growth by modifying its responses to light. When a crop plant is shaded by its neighbors, it tends to shoot upward to find the sunlight, spending energy that farmers want directed instead into making seeds. Certain proteins tell a plant when it's in the shade, and these proteins pass on the information to genes that control growth. By suppressing the activity of the shade-sensitive proteins, researchers can fool the plant into believing it is not in the shade and therefore has no reason to spend energy reaching skyward.

The American microbiologist Richard Jefferson, who discovered one of the early genetic marker genes, describes the function of the genes in a genome like the keys on a piano. "Imagine the keys of a piano. There are eighty-eight keys, and I know what each key means, but it doesn't tell me how to do Beethoven, Brahms, or Mozart. Yet all of that music is locked up in those keys. The secret is in their combinations, the order, the duration, and the intensity. It's the same way with genes."[22] Jefferson gives the example of teosinte, the very different-looking wild ancestor of maize.[23] Almost all the differences between the two are caused by only a few genes, and to a huge extent the difference in shape of the two plants is associated with just one single gene. "The key is how each gene regulates other genes," says Jefferson.

Many inspiring reports have emerged. One came from Norman Borlaug, the father of the Green Revolution in Mexico and Asia. In 2002 Borlaug, at eighty-eight, was reliving his earlier days as a plant breeder cultivating new varieties of corn in ten African countries, including Ethiopia, Uganda, Mozambique, and Ghana. With the help of funds from a fellow Nobel prizewinner, former President Jimmy Carter and his Atlanta-based Carter Center, Borlaug proudly declared that he could double or triple grain production in these ten countries within three years—if public funds could be found.[24]

At Cornell one of America's leading rice breeders, Susan McCouch, has been crossing commercial rice varieties with wild species and increasing yields by 10 to 20 percent. In some cases, McCouch has found her new varieties surprisingly resistant to rice plagues even though neither of the parents had such traits.

Researchers at Sussex University in England have produced salt-tolerant tomatoes without splicing a single gene. They found that tomatoes with the ability to tolerate the most salt in their tissues were the worst at stopping salt from entering their systems, while tomatoes bad at tolerating salt had the best methods of keeping it out. So they crossbred the two kinds. The results were tomatoes that were both good at preventing salt from entering their systems and good at tolerating salt should it pass their natural barrier.

Researchers are also looking for the genes that help a plant survive when it is under stress. They have found about two thousand genes that respond to various kinds of stress—exposure to salt, for example, or drought or low temperatures. Plants that don't deal well with stress possess the genes but for some reason don't switch them on. Researchers are trying to fix the wiring, as they say.[25] They believe that within a few decades they will be able to select and reactivate genes that cultivated plants used thousands of years ago when they were growing in much rougher habitats.

The study of genomes, or *genomics,* may help scientists find defenses against the potato blight mold that caused the destruction of Ireland's potato crops in the 1840s and still ravages potato fields around the world. The disease starts with purple-black lesions on the leaves; within a week it can turn the stalk and the potato itself to mush.

The fungus-like blight has recently turned up in Russia, destroying more of the country's staple crop than at any time in memory.[26] In countries that can afford it, the blight is treated with fungicide, but the disease is adept at mutating to survive even the poisons created to obliterate it. As an alternative defense, scientists have been studying blight-resistance genes in wild potatoes. One of these genes causes cells close to the infestation to, in effect, commit suicide, so that the mold cannot spread to other parts of the plant.

In the spring of 2002 two groups of researchers reported that they had mapped the entire genome for two types of rice, giving plant breeders an exciting new tool. Although the rice genome is the smallest of the major cereals—some six times smaller than corn and thirty-seven times smaller than wheat—the agbiotech companies focused on rice because of its similarity to other cereals and because it provides a rough guide to the possible location of useful genes in all major crops.

The genome for the *indica* variety was produced by China and the University of Washington, that of the *japonica* variety by the seed conglomerate Syngenta. Not all the new and important data have become publicly available. Researchers working on the *japonica* genome have had to sign a usage agreement with the Swiss-based company, protecting the information from Syngenta's competitors. Academic researchers can use the information for research, but not for commercial use. By contrast, the *indica* sequence has been put into a publicly available genome bank for use by rich corporations and poor researchers alike.

According to these genome maps, each rice cell contains forty to sixty thousand genes, compared to thirty to forty thousand genes in each human cell. Size isn't everything. The complexity of an organism does not depend on its gene count; it's how an organism uses those genes that matters. Animals have a system of generating a variety of different protein products from a limited number of genes. Scientists liken animal genes to a Swiss Army knife, one tool with lots of applications.

At his Canberra, Australia, nonprofit institute, Richard Jefferson directs research into what he calls *transgenomics*—a method halfway between traditional plant breeding and genetic engineering. Jefferson is effectively trying to mimic the natural process of evolution. He believes that there is a "Jurassic Park" of diversity slumbering inside the genome. The process does involve inserting artificially created genes—from yeast or bacteria—as the triggers or promoters, but for the purpose of generating new traits from the plant's own repertoire (to use Jefferson's piano analogy), not from alien genes. Once researchers have found ways to kick-start these genes in rice, corn should not be a problem. Botanists estimate that rice and corn began evolving from a common grassy ancestor at least 60 million years ago; their genomes are still largely identical. To be provocative, Jefferson is fond of saying, "Rice is corn," a comment deemed downright inflammatory in the Corn Belt of America. The differences arise when genes are switched on or off. Instead of moving a gene from an arctic flounder to a tomato to help the tomato survive a frost, Jefferson argues, it ought to be possible to goad a rice plant into a mutation or kick-start one of its own genes to produce a desired trait.

Scientists have known for decades that corn has its own self-help genes tucked away somewhere in its genome. Known as *transposons*, a kind of jumping gene, they often increase when a plant is under stress. "One of the genome's last-ditch responses under stress is to reshuffle the deck," Cornell's McCouch observes. It's a sort of panic response to find some way of dealing with the cause of the stress—heat, drought, cold, or plague.

In theory Jefferson's work presents an alternative for developing countries that cannot afford to use genes and techniques already patented by the seed congolmerates. Jefferson calls his process a way of "inventing around" those patents. Activist green groups generally applaud him. "It's a noble effort," says Hope Shand of ETC. But the activists are still concerned that corporate control over plant genomes will eventually result in greater economic control generally over the human food chain.

Even with Jefferson's transgenomics, the wide range of patents already owned still creates roadblocks for scientists trying to work for agriculture in developing countries. The more university researchers sign exclusive deals with biotech companies—such as the five-year deal Berkeley has with Novartis—the less those researchers are free to talk to other scientists. The change has been rapid. The small band of researchers looking into the asexual trait of apomixis have held two international conferences. At the first, in 1998, almost none of them had grants that tied them to corporations and there was a reasonably free exchange of ideas. At the second, in 2001, many of the researchers had become linked to corporate deals. As one of the participants put it, "No one gave anything away. They couldn't. They were all bought." [27]

To make biotechnology work to its full potential, especially in the developing world, researchers will need to use many of the genes and techniques owned by large corporations. There are hopeful signs that companies are willing to share part of their intellectual property with poor farmers. Syngenta's deal with Potrykus and Beyer over golden rice was an early example. Monsanto's transgenic sweet potato developed by Kenyan researchers was another. But Gary Toenniessen, the director of the Rockefeller Foundation's rice biotech program, who has watched the seed companies increase their control over lab research for thirty years, puts the new corporate image into perspective. Before making these rice genome data public, Monsanto and Syngenta selected what they wanted. "They've been mining the rice resource base as fast as they could," he says. "Despite all the rhetoric, these companies are still in the business to make money." [28]

To ensure that a flow of genetic information is available to developing countries, the Rockefeller Foundation has been trying to bring corporations and leading biotech universities together to create a pool of biotech tools—genes as well as laboratory techniques—that could be used free of royalties by researchers engaged in work specifically for poor countries. Others would like to see a radical restructuring of the patent system itself, perhaps one that would provide only limited protection for plant varieties, leaving genes and lab techniques free to be used for further research.

Until now I have deliberately avoided Malthusian discussions about biotech agriculture. Such debates mostly play straight into the hands of partisans. One side suggests that biotech is a magic bullet that will "feed the world"—in the memorable words of the Nobel prizewinning agronomist Norman Borlaug. The other side says that there is no such thing as a magic bullet for such a complex problem. There is already enough food in the world to keep people from starving. The solution, this side says, is not to produce more food but to enable people to afford what's available. The argument usually ends right there. It would be wrong, however, to leave a book about the future of food without taking the next step.[29]

In his 1798 *Essay on the Principle of Population,* the English clergyman Thomas Malthus foresaw an overcrowded future, a world with more people than food, resulting in global starvation and limits to the world's population. In those days, the earth's population was about 800 million. Today the population has grown to 6 billion, proving Malthus wrong, at least about upper limit to the number of human beings on the planet. During the nineteenth and twentieth centuries, with the help of new technologies, the amount of acreage cultivated across the globe became equal to the size of South America. We have doused those cleared fields with millions of tons of toxic chemicals in order to boost crop yields to meet the demand for food. While Malthus was wrong about the numbers, he was right about some people always being hungry. At the beginning of the twenty-first century, 800 million people still go to bed hungry every night.

By 2050 or so, the world population could start to level off at about 9 billion. Most of the increase in population will be in the developing world, where so many depend on rice as the basis of their diet. Yet rice yields have been stagnant for the last fifteen years in rice-producing countries such as Japan, Korea, and China. Increased yields have depended in the past on additional fertilizers, but in the fantastic yield increases of the Green Revolution, fertilizer is approaching its limit. Unused arable land is scarce, water supplies are dangerously low, and soils are impoverished. All the signs point to a future in which the world's poor will not be able to afford enough food to live on and in which the distribution system in undeveloped countries will not improve sufficiently to distribute food that is available.[30]

The question is, to what extent can biotechnology help in making up the shortfall? More than a decade ago, when young scientists were attracted to biotech agriculture, it promised two things: a reduction in the amount of pesticide use and the creation of an agriculture that would enable farmers, poor farmers especially, to produce food in a sustainable and more rational way. The new technology still holds out a possibility of cutting back on harmful chemicals, but the second promise has not turned out the way many had hoped. As one scientist from the early biotech era observes, "It got turned around on us. The technology became aligned with the corporate sector, with its objectives and its vision of the future. A lot of us who realized that it could service the different vision of sustainable agriculture are still struggling to give life to that alternative view."

Big corporations hijacked the technology, buying company after company primarily to expand their portfolios of biotech patents. Instead of inspiring inventors, as patents were originally intended to do, this intellectual property grab tended to exclude all but the rich corporate laboratories. Scientists had far less time, or inclination, to care about the public good, and those who saw biotech as a way to feed the world were hard to find. Vandana Shiva observed that the technology itself was seen as being "above society."[31]

There is no point in producing food that people refuse to eat. Nobody has learned this bitter lesson more thoroughly than biotech corporations and American farmers. The only way biotechnology can be switched back on track is by getting the public more involved. This means restoring the confidence of the grocery shopper by labeling GM foods. It means giving the public an opportunity to listen to reasoned arguments, no matter how complex the issue. With cloning and X-factoring and matrixing all part of public lore, most ordinary people already have some grasp of genes and genomes. So far, the public debate has been beneath them. Part of the blame rests with the companies who tried to sneak their products onto the market without telling people about them. Another part rests with those activist groups who took raising awareness beyond its usefulness and turned it into scaremongering.

At the beginning of 2004, the biotech industry was still strongly opposed in Europe and Japan. Monsanto shelved plans to market a GM wheat after North American farmers feared that they could not sell their crops in Europe. Britain completed the most extensive-ever field trials of three GM crops and concluded that the pesticides used with two of them—genetically modified sugar beet and oilseed rape—killed weeds more effectively, leaving fewer weeds to support wildlife—the key "quality of life" indicator being tested. The herbicide for a GM maize crop made by the German company Bayer was less damaging to the weeds, allowing more bees, butterflies, and birds to thrive. The British government approved cultivation of Bayer's maize, but also insisted on tough insurance regulations. Either the company or the farmers had to accept liability for any environmental damage caused by the crop. As farmers could not find insurance companies to underwrite the risk, the burden was on the company. Bayer withdrew its application to sell the corn. The industry's apparent victory turned sour.

More seemingly welcome news for the industry came from another front. The European Commission ended a six-year moratorium on the approval of new biotech crops and foods by allowing the Swiss biotech giant Syngenta to import a genetically modified sweet corn. However, the corn could only be imported, not grown in Europe, and the product would have to be clearly labeled "GM," under Europe's new rules. Green groups protested that Europe had been bullied into the approval after the United States had complained to the World Trade Organization that the moratorium represented unfair trading practice. The companies hoped that the strength of the opposition to GM crops would now be tested in the supermarket where market forces would hold sway; if the new GM corn was cheaper, then Europeans would buy it. But European retailers warned the companies not to expect a rush to buy; they even doubted that, given the anti-GM climate, the stores would stock the new corn.

Whatever the outcome of these battles in the already overfed capitals of Europe, they served the important function of raising again the crucial question about the use of the new technology—its potential in overcoming hunger in developing countries. The U.N.'s Food and Agriculture Organization, in its most sweeping endorsement yet of GM crops, declared that biotech agriculture held "clear promise" to alleviate global hunger and improve the lives of the poor. But the FAO lamented the lack of GM crops being made available in developing countries and called on both the private and public sectors in the U.S. and Europe to help those poor countries realize the full potential of the new technology. According to the FAO, biotech companies were spending ten times as much to improve crops for wealthy countries as governments and other donors were spending to improve crops for the poor. While it might be unrealistic to expect corporations to spend money on crops that brought them little return in capital, biotech research on the neglected "orphan crops" such as pearl millet, sorghum, and tef should be a public responsibility, said the FAO. The agency urged public funding of new biotech cash crops such as fruit and vegetables that are key to several developing nations' economies.

Predictably, the industry praised the FAO's report and the green groups dismissed it as more kowtowing to U.S pressure. Everyone agreed, however, that significant barriers to a change in the direction of GM research were still in place: the companies owned the patents to the lab techniques, as well as the products, and the developing countries in Africa and Asia lacked the technical expertise and the public funds to produce the type of seeds most suited to their needs.

At the same time, the biotech companies seemed to be consolidating their control rather than easing it. Monsanto had just won its patent suit against the Canadian canola farmer Percy Schmeiser. The suit had become a symbol of corporate patent control. After a bitter struggle, the Canadian Supreme Court rejected Schmeiser's explanation that some of Monsanto's GM seeds had landed on his farm by accident. The court ruled that any crop containing a patented seed, at whatever level, was covered by the provisions of that patent if the farmer tried to sell the crop. However, Schmeiser did not have to pay his profit of $14,000 on the crop to Monsanto. Monsanto hailed the ruling and Schmeiser's supporters said it was another reason for an urgent reform of the patent laws so that companies could not accumulate such power. Also in 2004, Monsanto started collecting royalties from Brazilian farmers who for several years had been growing contraband GM soybeans.

Despite prodding from the FAO, the idea of increasing public spending on "pro-poor" products also seemed a distant wish. The FAO noted that official bilateral development assistance to agriculture from OECD donor countries had actually fallen from $4.1 billion to $3.8 billion in 2002. The world's top ten transnational biotech companies were spending nearly $3 billion a year on agricultural biotech research and development. Private biotech research in most developing countries was negligible, and the developing countries with the largest public agricultural research programs—Brazil, China, and India—were spending less than half a billion dollars each. The largest international public supplier of agricultural technologies, the CGIAR, had a total annual budget for crop improvement of only about $300 million.

And yet against this dismal background a possible model for developing countries was emerging. China had been investing heavily in biotech, but with a very different approach. Instead of concentrating on the herbicide-resistant crops of the multinationals, China emphasized seeds that required lower levels of added chemicals—and had a "pro-poor" focus. Four GM crops have been commercialized—peppers, tomatoes, petunias, and cotton. In China, more than four million small farmers were growing insect-resistant cotton on about 30 percent of the country's total cotton area. Yields were reported to be about 20 percent higher than for conventional varieties; pesticide costs were around 70 percent lower, and pesticide use was reduced by a quarter. Foreign companies importing Bt cotton seeds were restricted to joint ventures whose access to local germ plasm and expansion was also strictly controlled. Local GM seeds were produced by the Chinese Academy of Agricultural Sciences—and were cheaper than foreign seeds. China had also formed its own company for exporting seeds to India, Vietnam, and Africa. The model was encouraging but not easily transferred to the smaller developing countries seeking to start up a biotech industry. For example, China and India, with strong domestic markets, were insulated from the effects of GM bans in Europe. But small African countries risked losing trade with Europe unless they remained "GM free."

For such countries, other trends in plant breeding seemed to hold out greater promise to "democratize" biotech agriculture—such as the techniques of researchers like Susan McCouch at Cornell, exploiting genomics to cross-breed rice with wild ancestors and produce offspring with traits their parents never showed, or Richard Jefferson's transgenomics, using genetic engineering to switch dormant genes on and off. Jefferson was also attracting attention—and public funds—for his ambition of an "open source" movement in biotechnology, like the open source movement in computer software, sharing the enabling tools of the trade to unleash the creativity of independent researchers, company laboratories, and farmers, all together. The idea harked back a century to the common public effort of breeding crops. He called his movement Biological Innovations for an Open Society, or BIOS, a name that certainly avoided the instant trepidation triggered by the term *genetically modified organisms,* or GMOs.

Biotech agriculture is another step in the evolution of human food, a process of change that began slowly and now, in evolutionary terms, moves at mach speed. The changes are not inherently unsafe, nor are the companies that produce them inherently evil. Transgenic foods have been eaten by contented and discerning consumers in America for a decade. Moreover, the promise of producing more food in African deserts or the wetlands of Asia is worth the time and money spent on these new seeds.

There are plenty of things for the public to worry about, however. One conern is how government agencies study and approve new seeds. In addition, old seeds must be preserved in public seed banks. Companies need to be more generous with patents that can be used to produce food in countries where people are starving. Genetic engineering has a pragmatic and realistic use for developing countries but only if it is properly integrated into the different agricultural systems. Finally, the strategic planners of world agriculture must bring an end to a system that through farm subsidies has long been rigged in favor of rich countries.

The glaringly unfair facts of agricultural subsidies and tariffs are worthy of constant repetition. Here are a few from a list produced by the International Food Policy Research Institute. If rich countries did not support their agriculture, developing countries could pick up annual gains of $40 billion, with sub-Saharan Africa, the world's poorest region, gaining $3.3 billion. Agricultural tariffs in the European Union and the United States are four to five times greater than those applied to manufactured goods. The $104 billion in subsidies that the EU provides its farmers accounts for one third of the value of the output, compared to one fifth in the United States. Beef and sugar are the most protected products in the EU and without these tariffs even the poorest African countries could produce exportable surpluses of these commodities. The EU accounts for 40 percent of the world's white sugar market, but the EU agricultural policy lowers the price of sugar by 15 percent. Despite modest attempts to reform that policy, EU subsidy spending will continue to rise for the next decade. Agricultural exports can help reduce poverty—they have played a critical role in reducing poverty in rural Uganda and Vietnam. The list goes on. In international negotiations the EU and the United States talk a lot about reforming their agricultural subsidies, but little gets done. Without meaningful reforms, poorer countries will not be able to compete in global agricultural markets.

The tinkering with genes in our food is not going to stop because some people consider the science a little freaky or believe that it has gone too far. Mendel's peas were revolutionary in their way in 1865. So were the modified tomatoes of 1992 and the golden rice of 2000, which may still help to prevent blindness in poor regions of the world. These GM groceries are not Frankenfoods any more than a person with a transplanted heart is today's Frankenstein. They are scientific creations full of both promise and potential hazard. These experimental foods deserve respect from those who discover them, call for more caution from those who regulate them and grow them, and finally, at the end of this real food chain, demand close study by those of us who eat them.

NOTES

Chapter 1. Mendel's Little Secret

1. Robin Marantz Henig, *The Monk in the Garden: The Lost and Found Genius of Gregor Mendel, the Father of Genetics* (Houghton Mifflin, 2000), 71–72.
2. Ibid., 171.
3. Menno Schilthuizen, *Frogs, Flies, and Dandelions: The Making of Species* (Oxford University Press, 2001), 29.
4. Jack Kloppenberg, *First the Seed: The Political Economy of Plant Biotechnology 1492–2000* (Cambridge University Press, 1998), 283, cited by Max John Pfeffer, "The Labor Process and Capitalist Development of Agriculture," *Rural Sociologist* 2, no. 2 (1982): 72–80.
5. Cary Fowler and Pat Mooney, *Shattering: Food, Politics, and the Loss of Genetic Diversity* (University of Arizona Press, 1990), 81.
6. Ibid., 66ff.
7. FAO, *The State of Food and Agriculture* annual review, 2000: 1.

Chapter 2. Seeds of Gold

1. *Time,* July 31, 2000, pp. 39–46.
2. Paul Brown, "GM Rice Promoters 'Have Gone Too Far,' " *Guardian* (London), February 10, 2001.
3. *Science,* January 14, 2000, 303.
4. Greenpeace called it "fool's gold," and GRAIN called it "grains of delusion."
5. Genetic Resources Action International (GRAIN), *Engineering Solutions to Malnutrition,* Barcelona, March 2000.
6. Charlie Kronick, Greenpeace U.K. Quoted by Dennis Avery in "Environmentalists Make Strong Case Against Enriched Rice," Knight-Ridder Tribune Business News, March 7, 2000.
7. GRAIN, ibid.
8. Brown, "GM Rice Promoters 'Have Gone Too Far.' " Also see Green-

peace, statement "Genetically Engineered 'Golden Rice' Is Fool's Gold,"
February 9, 2001.

9. Gordon Conway, Rockefeller Foundation, letter to Dr. Doug Parr of
Greenpeace (London), January 22, 2001.

10. International Food Policy Research Institute (IFPRI), *Vitamin A Problems of Feeding UDCs,* 2002, a review, specifically chapters 4 and 5.

11. Bill Lambrecht, *Dinner at the New Gene Café: How Genetic Engineering Is Changing What We Eat, How We Live, and the Global Politics of Food* (St. Martin's Press, 2001), 282–85.

12. Kathleen Hart, *Eating in the Dark: America's Experiment with Genetically Engineered Food* (Pantheon Books, 2002), 215.

13. E. J. Kahn, *The Staffs of Life* (Little, Brown, 1984), 214.

14. Ingo Potrykus, "Golden Rice and Beyond," *Plant Physiology,* 125 (March 2001): 1157–61.

15. Author interview, Zurich, March 22, 2001.

16. Potrykus, letter to Hope Shand of Rural Advancement Foundation International (RAFI), October 18, 2000.

17. Daniel Charles, *Lords of the Harvest: Biotech, Big Money, and the Future of Food* (Perseus, 2001), 85–86.

18. Author interview, Zurich, March 22, 2001.

19. *Agrobacterium tumefaciens* was first named by researchers from the U.S. Department of Agriculture in 1907.

20. Author e-mail interviews with Potrykus, 16 and 22 July 2002. Xudong Ye, et al., "Engineering Pro-Vitamin A (Beta-carotene) Biosynthetic Pathway into (Carotenoid-free) Rice Endosperm." *Science* 287 (2000): 303–305.

21. The funds came from the Carotene Plus project of the EU, of which golden rice would become a small part. The EU rules—Framework IV and V—require public research institutions, such as Bayer's, to take a partner from industry and for that partner to have a share in the result.

22. Author interview, Freiburg, October 25, 2001.

23. Author interview, March 22, 2001.

24. International Service for the Acquisition of Agri-biotech Applications (ISAAA) Brief: The Intellectual and Technical Property Components of Pro-Vitamin Rice: A Preliminary Freedom to Operate Review, no. 20 (2000).

25. Potrykus, "Golden Rice and Beyond."

26. "Golden Goosed? Update on Trojan Trade Reps, Golden Rice, and the Search for Higher Ground," press release from RAFI, October 12, 2000.

27. GRAIN, Engineering Solutions.

Chapter 3. The Plague of Sameness

1. Otto and Dorothy Solbrig, *So Shall You Reap* (Island Press, 1994).
2. RAFI, 1982, cited in Fowler and Mooney, *Shattering,* chapter 4.
3. Paul Raeburn, *The Last Harvest: The Genetic Gamble That Threatens to Destroy American Agriculture* (Simon & Schuster, 1995), 93.
4. Penelope Francks, *Technology and Agricultural Development in Pre-war Japan* (Yale University Press, 1984), 48.
5. Kahn, *Staffs of Life,* 63.
6. Fowler and Mooney, *Shattering,* 81.
7. J. Harlan and M. L. Martini, "Problems and Results in Barley Breeding," *Yearbook of Agriculture, 1936,* U.S. Government Printing Office.
8. Kloppenberg, *First the Seed,* 158.
9. The 1913 writ of the Rockefeller Foundation included promoting "the well-being of mankind throughout the world." The foundation had taken on hunger as part of that goal.
10. William Gaud, administrator of USAID.
11. Donald Duvick, "Biotechnology in the 1930s: The Development of Hybrid Maize." *Nature Reviews/Genetics* 2 (January 2001), 69–74, and author interview.
12. Kloppenberg, *First the Seed,* 94.
13. National Academy of Sciences, *Genetic Vulnerability of Major Crops* (National Academy Press, 1972).
14. FAO, *The State of Food and Agriculture.*
15. There were seven firms producing ammonia (the basis of much nitrogen fertilizer) in 1940; by 1966 there were sixty-five. By 1980, eight companies owned more than 70 percent of the seed corn market.
16. Kloppenberg, *First the Seed,* 117.
17. Gordon Conway, *The Doubly Green Revolution: Food for All in the Twenty-first Century* (Cornell University Press, 1997), 49.
18. Raeburn, *The Last Harvest,* 93.
19. Andrew Pearse, *Seeds of Plenty, Seeds of Want: Social and Economic Implications of the Green Revolution* (Clarendon Press, 1980), 37.
20. Ibid., 57.
21. R. Pistorius and J. van Wijk, *The Exploitation of Plant Genetic Information: Political Strategies in Crop Development* (CABI Publishing, 1999), 97.
22. Per Pinstrup-Andersen and Ebber Schiøler, *Seeds of Contention: World Hunger and the Global Controversy Over GM Crops* (Johns Hopkins, 2000).

23. Unknown to the West at the time, China started using dwarf varieties of rice in 1959, with spectacular results. It was later discovered that China was using the same dwarfing gene variety as Taiwan. China also introduced the new wheat strains from Mexico, crossing them with local varieties.

24. Frances Moore Lappé, Joseph Collins, and Peter Rosset, with Luis Esparza. *World Hunger: Twelve Myths, Food First* (Grove Press, 1998), 61.

25. M. S. Swaminathan, New Agriculturist: Perspective, The Challenges Ahead 99–4. 1999; http://www.new-agri.co.uk/99-4/perspect.html.

26. N. Alexandratos, *World Agriculture: Towards 2010,* FAO Study (Chichester [UK]: Wiley & Sons, 1995), also quoted in Conway, *Doubly Green,* 87.

27. Conway, *Doubly Green,* 212.

28. Ibid., 214–15.

29. Ibid., 59.

30. Pearse, *Seeds of Plenty,* 26.

31. Per Pinstrup-Andersen, 18.

32. Lappé et al., *World Hunger,* 76.

33. See Conway, *Doubly Green.*

Chapter 4. A New Sort of Tomato

1. Michael Hansen, Genetic Engineering Is Not an Extension of Conventional Plant Breeding: How Genetic Engineering Differs from Conventional Breeding, Hybridization, Wide crosses, and Horizontal Gene Transfer, Consumer Policy Institute/Consumers Union, January 2000. 1–14.

2. Alan McHughen, *Pandora's Picnic Basket: The Potential Hazards of Genetically Modified Foods* (Oxford University Press, 2000), 187–89.

3. *Introduction of Recombinant DNA-Engineered Organisms into the Environment: Key Issues,* National Academy of Sciences (NAS), 1987.

4. Kurt Eichenwald, Gina Kolata, and Melody Peterson, "Biotechnology Food: From Lab to a Debacle," *New York Times,* January 25, 2001.

5. Ibid., quote from Earle Harbison, Monsanto president and chief operating officer in late 1980s.

6. Eric Millstone, Eric Brunner, and Sue Mayer, "Beyond 'Substantial Equivalence,' " *Nature,* October 7, 1999, 525–26.

7. *Four Principles of Regulatory Review: 1992,* National Biotechnology Policy Board Report, National Institutes of Health, E-6; "Statement of Food Policy: Foods Derived from New Plant Varieties"; and *Federal Register.* For an exhaustive overview of the origins of the U.S. regulatory

system of genetically engineered foods, see Consumer Federation of America Foundation's report, *Breeding Distrust: An Assessment and Recommendations for Improving the Regulation of Plant Derived Genetically Modified Food*, February 2001, especially chapters 3 and 4. Also see Hart, *Eating in the Dark*, 78–9.

8. The comments of FDA scientists were made public in 1999 as a result of a lawsuit brought by the Alliance for Bio-Integrity et al. v. Donna Shalala et al. See also Marion Burros, "Documents Show Officials Disagreed on Altered Food," *New York Times*, December 1, 1999.

9. David Kessler, "Companies are now ready," memorandum to Secretary of Health and Human Services, "FDA Proposed Statement of Policy Clarifying the Regulations of Food from Genetically Modified Plants, Decision, March 20, 1992," quoted in Hart, *Eating in the Dark*, 81.

10. "Statement of Food Policy," quoted in Hart, *Eating in the Dark*, 79.

11. Eichenwald, "Biotechnology Food."

12. Millstone et al., "Beyond 'Substantial Equivalence.' "

13. Liam Donaldson and Sir Robert May, *Health Implications of Genetically Modified Foods*. Ministerial Group on Biotechnology (UK), May 1999.

14. Ibid., 182.

15. Millstone et al., "Beyond 'Substantial Equivalence.' "

16. Introduction of Transgenes into Plants: Genetically Modified Pest-Protected Plants: Science and Regulation, U.S. Material Academy of Science, National Research Council, April 2000.

17. Stephen Nottingham, *Eat Your Genes* (Zed Books, 1998), 76. See also Martin Teitel and Kimberly Wilson, *Genetically Engineered Food: Changing the Nature of Nature* (Rochester, Vermont: Park Street Press, 2001), 25.

18. Belinda Martineau, *First Fruit: The Creation of the Flavr Savr Tomato and the Birth of Genetically Engineered Food*, 58.

19. Ibid., 113–14.

20. Ibid., 68 and 113–17.

21. McHughen, *Pandora's Picnic Basket*, 89.

22. Martineau, *First Fruit*, 117.

23. Consumer Federation of America Foundation Report, February 2001 op. cit. Chapter 1:18., also Pew Initiative on Food and Biotechnology: Guide to U.S. Regulation of Genetically Modified Food and Agricultural Biotechnology Products. September 7, 2000.

24. Ibid.

25. Martineau, *First Fruit*, 84.

26. Ibid., 67.

27. Ibid., 65.

28. Ibid., 83.

29. A. Barnum, "Biotech Tomato Wins Final OK for Marketing," *San Francisco Chronicle,* May 19, 1994.

30. W. Leary, "FDA Approves Altered Tomato That Will Remain Fresh Longer," *New York Times,* May 19, 1994.

31. Barnum, "Biotech Tomato."

32. "Novartis Unveils New Gene Marker," Reuters, May 23, 2000.

33. Hart, *Eating in the Dark,* 296.

Chapter 5. The Battle of Basmati

1. V. P. Singh, "The Basmati Rice of India," in *Aromatic Rices,* R. K. Singh, U. S. Singh, and G. S. Khush, eds. (Science Publishers, 2000), 135.

2. R. K. Singh et al., *Aromatic Rices,* 164.

3. *Down to Earth Magazine,* Centre for Science and Environment, India 2000. http://www.cseindia/html/dte/dte20011015/dte_life.htm.

4. RAFI invented the term *biopiracy* to refer to uncompensated commercial use of biological resources or traditional knowledge from developing countries.

5. R. K. Singh et al., *Aromatic Rices,* 41.

6. Ibid., 52.

7. Jasmine is low in amylose, it's sticky, and its grains remain the same length in the rice pot.

8. RAFI discovered RiceTec's exotic past: the company's owner turned out to be Lichtenstein's royal family, headed by reigning Prince Hans-Adam II. His personal fortune, they noted, was somewhere between $3.75 and $4.25 billion. In May 1998 the greens, led by RAFI, launched a postcard campaign against the granting of the patent. Despite receiving thousands of petitions that he abandon the patent, Prince Hans-Adam refused to budge. The green groups went to see the prince to try to convince him that RiceTec did not invent basmati rice, the poor hill farmers of India and Pakistan did. But the prince was not impressed. RAFI, Basmati Rice Update, January 4, 2000.

9. *Human Development Report,* United Nations Development Program (UNDP), (Oxford University Press, 1999), chapter 2, 66–76.

10. John Vidal, *Guardian.* November 16, 2000.

11. "UN Moves to Curb Biopiracy," BBC report from The Hague, April 17, 2002.

12. J. Maddley, Yours for Food, report commissioned by Christian Aid (UK), 1996. Quoted in Luke Anderson, op. cit.: 83.

13. Kloppenberg, 152. Plants and their genetic resources had always been considered the "common heritage of mankind," and the wrangling over the Law of the Sea treaty highlighted the problem of the developed countries' giving up "common heritage" status on resources outside their national boundaries.

14. Nigel Dower, *Biotechnology and the Third World* (University of Aberdeen, 2000). http://www.abdn.ac.uk/~phl002/a4.htm.

15. Subsequent U.S. patent statutes would employ the same broad language, ending with a 1952 amendment that included protection for "anything under the sun that is made by man."

16. Kloppenberg, *First the Seed,* 65.

17. Ibid., 132.

18. Calestous Juma, *The Gene Hunters: Biotechnology and the Scramble for Seeds* (Princeton University Press, 1989), 159.

19. *Colorado Journal of International Environmental Law and Policy,* Plants and Politics 7, no. 1 (1995): 122–26.

20. Ibid.

21. Brian Belcher and Geoffrey Hawtin, *A Patent on Life: Ownership of Plant and Animal Research* (IDRC, 1991), quoted in Anderson, *Genetic Engineering,* 70.

22. Mark Sagoff, "Patented Genes: An Ethical Appraisal," Issues in Science and Technology Online, Spring 1998. http://www.nap.edu/issues/14.3/sagoff.htm.

23. Jeremy Rifkin, *The Biotech Century* (Tarcher and Putnam, 1998), 43.

24. Kloppenberg, *First the Seed,* 267.

25. Sagoff, "Patented Genes."

26. Ibid.

27. Brian Wright, "Intellectual Property Rights Challenges and International Research Collaborations in Agricultural Technology," in Qaim et al., eds., *Agricultural Biotechnology in Developing Countries: Towards Optimizing the Benefits for the Poor* (Netherlands: Kluwer Academic Publishers, 2000), 289–314.

28. Ibid.

29. W. Lambert and A. S. Hayes, "Investing in Patents to File Suits Is Curbed," *Wall Street Journal,* May 30, 1990.

30. Kristi Coale, "The Contract and the Code: When Corporations Own Knowledge," University Business, October 1999. http://web.mitretek.org/rockefeller/biote . . . 8007461da?openDocument& Highlight-2, tomato.

31. William Lesser, *Intellectual Property Rights and Concentration in*

Agricultural Biotechnology (Cornell University, 1998, 1–6). http://www. agbioforum.org/vol1no2/lesser.htm.

32. GRAIN, Patenting, Piracy and Perverted Promises, 1998; RAFI, Bioprospecting/Biopiracy and Indigenous Peoples; F. Powledge, "Who Owns Rice and Beans?" *BioScience,* July/August 1995, 440–44; Anderson, *Genetic Engineering,* 71.

33. RAFI, *Bioprospecting,* 26.

34. Barton and Berger, Patenting Agriculture, Issues in Science and Technology Online, Summer 2001, 1–12.

35. Ibid., 1–12.

Chapter 6. Of Cauliflower, Potatoes, and Snowdrops

1. Alan Davidson, *The Oxford Companion to Food* (Oxford University Press, 1999), 147.

2. Mae-Wan Ho and Angela Ryan, "Hazards of Transgenic Plants Containing the Cauliflower Mosaic Viral Promoter," Institute of Science in Society, and Biology Department, Open University, UK.

3. Mae-Wan Ho, "Special Safety Concerns of Transgenic Agriculture and Related Issues," Institute of Science in Society, *ISIS News* 3 (1999).

4. Roger Hull, S. Covey, and P. Dale, "Genetically Modified Plants and the 35S Promoter: Assessing the Risks and Enhancing the Debate," *Journal of Microbiology, Ecology, Health, and Disease* 12 (2000): 1–5.

5. Ho and Ryan, "Hazards of Transgenic Plants."

6. Jean-Benoit Morel and Mark Tepfer, "Are There Potential Risks Associated with the Cauliflower Mosaic Virus 35S Promoter in Transgenic Plants?" Laboratoire de Biologie Cellulaire, France, 2000. http://www. biotech-info.net/CMV_risks.html.

7. John Hodgson, "Scientists Avert New GMO Crisis," *Nature Biotechnology* 18, no. 1 (January 2000): 13.

8. Ibid.

9. Morel and Tepfer, "Potential Risks."

10. Roger Hull et al., "Genetically Modified Plants and the 35S Promoter."

11. "Genetically Modified Plants for Food Use and Human Health—an Update," Royal Society, London, February 2002.

12. "The Golden Rice—An Exercise in How Not to Do Science," Institute of Science in Society. http://www.i-sis.org/rice.shtml.

13. Julian Borger, "How the Mighty Fall," *Guardian,* November 22, 1999.

14. Yousef Ibrahim, "Genetic Soybeans from the U.S. Alarm Europeans," *New York Times,* November 7, 1996.

15. Diane Osgood, *Dig It Up: Global Civil Society's Responses to Plant Biotech-*

nology, Global Civil Society Year Book 2001 (Oxford University Press), 79–107.

16. Charles, *Lords of the Harvest,* 206.

17. Julian Kinderlerer, *Public Reaction to Genetically-Modified Foods in the UK,* Sheffield Institute of Biotechnological Law and Ethics, and Department of Law, University of Sheffield, 2000, a paper for the Center for International Development at Harvard University.

18. The supermarkets were J. Sainsbury and Marks and Spencer in the U.K., Carrefour in France, Migros in Switzerland, Delhaize in Belgium, Superquinn in Ireland, and Effelunga in Italy.

19. Maria Margaronis, "The Politics of Food," *The Nation,* December 27, 1999, 11–16.

20. Ibid.

21. Charles, *Lords of the Harvest,* 217.

22. Anderson, *Genetic Engineering,* 49.

23. McHughen, *Pandora's Picnic Basket,* 153.

24. Ibid., 8.

25. Ibid., 7.

26. World in Action TV program, UK, How Safe Is Genetically Modified Food? August 8, 1998.

27. W. P. T. James and James Chesson, *The Scientific Advisory System: Genetically Modified Food Inquiry,* Rowett Research Institute, Aberdeen, Scotland, 1 March 1999, 14.

28. "GMO Food," Greenpeace press release, February 12, 1999.

29. W. B. Stanley and A. Pusztai, "Effects of Diets Containing Genetically Modified Potatoes Expressing *Galanthis rivalis* Lectin on Rat Small Intestine," *The Lancet* 354, no. 9187 (October 16, 1999): 1353.

30. Richard Horton, "Genetically Modified Foods: Absurd Concern or Welcome Dialogue," *The Lancet* 354, no. 9187 (October 16, 1999): 1312.

31. Interview with Horton by *Guardian* journalists, Laurie Flynn and Michael Sean Gillard, "Pro-GM food scientist 'threatened editor,' " October 21, 1999.

32. Horton, "Genetically Modified Foods."

33. H. A. Kulper et al., "Adequacy of Methods for Testing the Safety of Genetically Modified Foods," *The Lancet* 354, no. 9187 (October 16, 1999): 1313.

34. Charles, 210.

35. "Let Prince Charles Travel by Bullock Cart When He Comes to India," *LM Magazine Online,* June 12, 2000. http://mitretek.org/Rockefeller/Blote . . .

36. Richard Dawkins, "Your Highness, Science Is Not Perfect, But Nor Is Nature: In an Open Letter to Prince Charles, Richard Dawkins Calls for Reason," *Ottawa Citizen,* June 1, 2000, A19.

37. Andrew Marr, "Charles: Right or Wrong about Science?" The *Observer,* London, May 21, 2000.

38. "Church Ban on GM Crop Trials," *Independent,* Aug 4, 1999.

39. Pontifical Academy for Life, Vatican City, October 12, 1999. Report on http://www.agbioworld.org/articles/vatican.html.

40. Scott Kilman and Thomas Burton, "Biotech Backlash Is Battering Plan Shapiro Thought Was Enlightened," *Wall Street Journal,* December 21, 1999, A1.

41. Michael Specter, "The Pharmageddon Riddle," *New Yorker,* April 10, 2000, 58–71.

Chapter 7. Anatomy of a Poisoned Butterfly

1. A. M. Shelton, J. Z. Zhao, and R. T. Roush, "Economic, Ecological, Food Safety, and Social Consequences of the Deployment of Bt Transgenic Plants," *Annual Review of Entomology* 47 (2002): 845–81.

2. Karen Oberhauser et al., "Temporal and Spatial Overlap Between Monarch Larvae and Corn Pollen," *Proceedings of the National Academy of Sciences,* October 9, 2001: 1–13.

3. A. F. Krattiger, "Insect Resistance in Crops: A Case Study of *Bacillus thuringiensis* (Bt) and Its Transfer to Developing Countries," *ISAA Briefs* 2 (1997): 42. Also quoted in Shelton et al., "Consequences."

4. Eric Grace, *The World of the Monarch Butterfly* (Sierra Club Books, 1997), 61.

5. Ibid., 12.

6. Charles, *Lords of the Harvest,* 245.

7. Cornell University press release, September 10, 1999.

8. Author interview, October 26, 2002.

9. J. Losey, L. Raynor, and M. E. Carter, "Transgenic Pollen Harms Monarch Larvae," *Nature* 399 (May 20, 1999), 214.

10. "Pesticide Fact Sheet for *Bacillus thuringiensis,*" 1995, Environmental Protection Agency Publ. EPA731-F-95–004.

11. Shelton et al., *Consequences.*

12. Pew Initiative on Food and Biotechnology, Three Years Later: Lessons Learned from the Monarch Butterfly Controversy, May 30, 2002, quoting Janet Andersen, director of the Biopesticides and Pollution Prevention Division at EPA, 3.

13. Ibid.
14. "Monarch Butterflies and Toxic Pollen," Union of Concerned Scientists (UCS). http://www.ucsusa.org/food/monarch.html.
15. Anthony Shelton, Comments on Recent Reports Involving Plants Engineered to Protect Them from Insect Attack, Congressional Testimony, October 5, 1999.
16. Richard Shelton and Richard Roush, "Two Leading Researchers Take Issue with Recent Studies on the Effects of Genetically Modified Crops," Cornell University press release, September 10, 1999.
17. Monsanto PR Newswire, Monsanto statement on Bt corn: Environmental Safety and a Recent Report on the Monarch Butterfly, May 20, 1999.
18. Agricultural Research Service, USDA, Research Q&A: Bt Corn and Monarch Butterflies, 2001 Summary of Proceedings of the National Academy of Sciences articles on risk to the monarch butterfly.
19. Author interview. October 26, 2002.
20. Author interview. October 26, 2002.
21. "Monarch Butterflies and Toxic Pollen."
22. L. C. Hansen and J. J. Obrycki, "Field Deposition of Bt Transgenic Corn Pollen: Lethal Effects on the Monarch Butterfly," *Oecologia* 125 (2000): 241–48.
23. Kevin Steffey, "Here We Go Again: Bt and Corn Monarch Butterflies," AgBioView, August 26, 2000. http://agbioview.listbot.com/cgi-bin/subs . . . st_id=agbioview&msg_num=761&start_num-761.
24. Pew, Lessons Learned from Monarch Butterfly Controversy: 8–12.
25. Author interview with Losey, October 26, 2000.
26. Pew, Lessons Learned, 8.
27. Diane Stanley-Horn et al., "Assessing the Impact of Cry-1Ab-expressing Corn Pollen on Monarch Butterfly Larvae in Field Studies," *Proceedings of the National Academy of Sciences* 10.1073/pnas.21127798. September 14, 2001, 1–14.
28. Rebecca Goldburg, preliminary research results presented during the Monarch Research Symposium, Environmental Defense Fund, November 2, 1999.
29. Pew, Lessons Learned, quote from University of Guelph researcher, Mark Sears: 12.
30. Andrew Pollack, *New York Times,* September 8, 2001.
31. Pew, Lessons Learned, 12.
32. Pew, Lessons Learned, 14.
33. Pew, Lessons Learned, quoting EPA's Anderson, 15.
34. Pew, Lessons Learned, quoting Hemlich and Mellon, 17.

Chapter 8. The Plant Hunters

1. Nikolai Vavilov, *Five Continents,* ed. L. E. Rodin, translated from the Russian by Doris Love, with ed. Seymon Reznik and Paul Stapleton. (International Plant Genetic Resources Institute, Rome/VIR, St. Petersburg), 1997, 102–107.

2. *Dictionary of Scientific Biography,* ed. C. C. Gillespie, vol. 15, Supplement 1 (Charles Scribner's Sons, 1978), 509.

3. Raeburn, *Last Harvest,* 42.

4. Chauncy Harris, *N. I. Vavilov 1887–1943, Working Groups on the History of Geographical Thought of the International Geographical Union and the International Union of the History of Philosophy of Science,* ed. G. J. Martic (London: Mansell, 1991), 117–32.

5. M. Flintner, Sammler, Rauber und Gelehrte: *Die Politischen Interessen an Pflanzengenetischen Ressourcen 1895–1995* (Frankfurt, Germany: Campus Verlag, 1995), quoted in Pistorius and van Wijk, *Exploitation,* 62.

6. Vavilov, *Five Continents,* xix.

7. Ibid., xxvii.

8. Author interview, November 11, 2001.

9. RAFI interview with Professor Kelly, in Mexican Bean Conspiracy, U.S.-Mexico Legal Battle Erupts Over Patented Enola Bean, Plant Breeders' Wrongs Continue, RAFI briefing, January 17, 2000.

10. Timothy Pratt, "Patent on Small Yellow Bean Provokes Cry of Biopiracy," *New York Times,* March 20, 2001.

11. Ibid.

12. Joel Cohen and Robert Paarlberg, "Explaining Restricted Approval and Availability of GM Crops in Developing Countries," *AgBiotechNet* 4, October 2002.

13. Ibid.

14. Calestous Juma, "Biotechnology and International Relations: Forging New Strategic Partnerships," draft paper to *International Journal of Biotechnology,* 2001.

15. Richard Lewontin, "Genes in the Food," *New York Review of Books,* June 21, 2001, 81–84.

16. Ibid.

17. *Economist,* March 27, 2000.

18. Jacques Diouf, FAO at the World Agricultural Forum 2001.

19. Hans Herren, Director General, International Center of Insect Physiology and Ecology, Nairobi, Kenya, see "Expert Doubts Africa's Gain from Genetically Modified Crops," *The Nation,* Nairobi, June 3, 1999. See

also paper by Miguel Altieri and Peter Rossett, "Ten Reasons Why Biotechnology Will Not Ensure Food Security, Protect the Environment, and Reduce Poverty in the Developing World," 2000. http://www.iatp.org.foodsec/library/admin.htm.

20. Vandana Shiva, "Biopiracy—U.S. Patent Law Must Change, *The Hindu,* Jul. 28, 1999.

21. "TRIPS versus CBD: Conflicts Between WTO Regime of Intellectual Property Rights and Sustainable Biodiversity Development," GRAIN, April 1998.

22. For a discussion of pros and cons for international property rights in developing countries, see Robert Herdt, *Enclosing the Global Plant Genetic Commons,* Institute for International Studies, Stanford University, 1999.

23. The twelve nations, known as the Group of Allied Mega-Biodiverse Nations, was formed at Cancún, Mexico, in February 2002.

24. Juma, "Biotechnology and International Relations."

25. J. DeVries and G. Toenniessen, *Securing the Harvest: Biotechnology, Breeding, and Seed Systems for African Crops* (CABI Publishing, 2001), 20–21.

26. Devlin Kuyek, "Intellectual Property Rights in African Agriculture," GRAIN, August 2002.

27. Ibid.

28. Barbara Woodward, Johan Brink, and Dave Berger, "Can Agricultural Biotechnology Make a Difference in Africa?" ARC-Roodeplaat Vegetable and Ornamental Plant Institute, Pretoria, South Africa, 2001. http://www.agbioforum.org/vol2no34/woodward.htm.

29. Interview with Florence Wambugu, *New Scientist,* May 27, 2000; also Wambugu, "Why Africa Needs Agricultural Biotech," *Nature,* July 1, 1999.

30. Author interview, November 2, 2001.

31. Marilyn Berlin Snell. *Against the Grain: Why Poor Nations Would Lose in a Biotech War on Hunger* (Sierra Club, 2002).

Chapter 9. The Cornfields of Oaxaca

1. Charles Clover, " 'Worst Ever' GM Crop Invasion," *The Hague Daily Telegraph,* April 19, 2002.

2. Alan Zarembo, "The Tale of the Mystery Corn in Mexico's Hills," *Newsweek,* January 28, 2002, 41 (international edition).

3. Matt Metz, press release on transgenic corn in Mexico, Dept. of Microbiology, University of Washington, Seattle, April 5, 2002.

4. Marc Kaufman, "Journal Editors Disavow Article on Biotech Corn," *Washington Post,* April 4, 2002 A3.
5. Kara Platoni, "Kernels of Truth," *East Bay Express,* May 29, 2002. http://www.biotech-info.net/kernels of truth.html.
6. Luis Herrera Estrella, "No Evidence That Transgenic Maize in Oaxaca Is a Threat to Biodiversity in Mexico," CINVESTAV Unidad Irapuato, www.checkbiotech.org, December 4, 2001.
7. Zarembo, "The Tale of the Mystery Corn"; also Mark Shapiro, "Sowing Disaster? How Genetically Engineered American Corn Has Altered the Global Landscape," *The Nation,* October 28, 2002, 11–19.
8. Alan Zarembo, "The Tale of the Mystery Corn in Mexico's Hills."
9. Author interviews with Ignacio Chapela, November 27 and 29, 2002.
10. "Bt Gene Flow in Mexico: Preserving the Integrity of Indian Corn, Part 1," *Indian Country News,* February 8, 2002.
11. David Quist and Ignacio Chapela, "Transgenic DNA Introgressed into Traditional Landraces in Oaxaca, Mexico," *Nature* 414, no. 6863 (2001): 541–43.
12. Marc Kaufman, "Journal Editors Disavow Article on Biotech Corn," *Washington Post,* April 4, 2002, A03.
13. CIMMYT, Review of Oaxaca controversy, May 8, 2002. Also see Mauricio Bellon and Julien Berthaud, In-situ Conservation of Maize Diversity, Gene Flow and Transgenes in Mexico, OECD Conference, Raleigh-Durham. November 27–30, 2001.
14. Platoni, "Kernels of Truth."
15. Kaufman, "Journal Editors."
16. Nick Kaplinsky, "Biodiversity (Communications Arising): Maize Transgene Results in Mexico Are Artifacts," *Nature,* April 4, 2002, 739.
17. Matthew Metz and Johannes Futterer, "Biodiversity (Communications Arising): Suspect Evidence of Transgenic Contamination," *Nature,* April 4, 2002, 738.
18. David Quist and Ignacio Chapela, "Biodiversity (Communications Arising): Suspect Evidence of Transgenic Contamination/Maize Transgene Results in Mexico Are Artifacts," *Nature,* April 4, 2002, 740.
19. Ignacio Chapela, letter to *Guardian* (London), May 24, 2002.
20. Ibid.
21. ABC Science Program, *The Lab on Line,* July 2002. http://www.life sciencesnetwork.com/news-detail.asp?new ID=1708.
22. Fred Pearce, "The Great Mexican Maize Scandal," *New Scientist,* June 15, 2002. http://www.biotech-info.net/special.html.

23. Ibid., 166.

24. Ibid.

25. "U.S. Scientists Defend Mexico Corn GMO Study," Reuters, April 4, 2002.

26. Norman Ellstrand, "Evaluating the Risks of Transgene Flow from Crops to Wild Species," Mexican government seminar organized by CIMMYT, 1995; "Gene Flow Among Maize Landraces, Improved Varieties and Teosinte: Implications for Transgenic Maize, the Mexican Institute of Forestry, Agriculture and Livestock Research, and the Mexican National Agricultural Committee, Mexico, 1995.

27. "Gene Flow among Maize Landraces, Improved Maize Varieties, and Teosinte: Implications of Transgenic Maize," op. cit.

28. ETC Group Communiqué "Fear Reviewed Science," January/February 2002.

29. Norman Ellstrand, "When Transgenes Wander, Should We Worry?" *Plant Physiology* 125(April 2001): 1543–45, quoting from R. M. Goodman and N. Newell, "Genetic Engineering of Plants for Herbicide Resistance: Status and Prospects," in H. O. Halvorson, D. Pramer, M. Rogul, eds., *Engineered Organisms in the Environment: Scientific Issues* (American Society for Microbiology, 1985), 47–53.

30. Jane Rissler and Margaret Mellon, *The Ecological Risks of Engineered Crops* (MIT Press, 2000), 56.

31. Ellstrand, "Evaluating the Risks of Transgene Flow."

32. Ibid.

33. Rissler and Mellon, *Ecological Risks*, 56.

34. John Jemison and Michael Vayda, "Cross-Pollination from Genetically Engineered Corn: Wind Transport and Seed Science," University of Maine, *AgBioForum* 4, no. 2: 87–92.

35. Rissler and Mellon, *Ecological Risks*, 52.

36. AgBioWorld: Joint Statement in Support of Scientific Discourse in Mexicans GM Maize Scandal, February 2002. http://www.agbioworld.org.jointstatement.html.

37. Author interviews, Mauricio Bellon and Julien Berthaud, International Maize and Wheat Improvement Center, Mexico, November 4, 2002.

38. *Morning Edition*, National Public Radio, December 18, 2001.

39. CIMMYT, Review of Oaxaca Controversy, op. cit., May 8, 2002.

40. Monsanto UK, background information following Canadian biotech court case decision, April 18, 2001.

41. Charles, *Lords of the Harvest*, 187.

42. McHughen, *Pandora's Picnic Basket,* 164.

43. Murray Lyons, "Farmer's Reapings No Fluke, Court Told," *Saskatoon Star Phoenix,* June 6, 2000.

44. Ed White, "Farmer's Story Lacks Credibility, Says Scientist," *Western Producer,* Saskatoon, June 15, 2000.

45. McHughen, *Pandora's Picnic Basket,* 165.

46. Martin Phillipson, "Commentary: Monsanto v. Schmeiser," "Bar Notes," Canadian Bar Association, Saskatchewan Branch, June 2001. http://www.biotech-info.net/phillipson-commentary.html.

47. Ellstrand, "When Transgenes Wander."

Chapter 10. So Shall We Reap

1. Richard Ragan, WFP representative quoted by PanAfrica News Agency, August 13, 2002.

2. The Biosafety Protocol comes into force when fifty nations have ratified it. At the time of writing, thirty-seven had done so, including the fifteen nations of the EU, several African nations, and India, New Zealand, Austria, and Denmark.

3. Devlin Kuyek, "Genetically Modified Crops in African Culture," GRAIN, August 2002. Reference from R. Paarlberg, "Policies Towards GM Crops in Kenya," in *Governing the GM Crop Revolution: Policy Choices for Developing Countries.* 2020 Vision Food, Agriculture and the Environment; discussion paper 33, December 2000. See also Kristin Dawkins, "Biotech from Seattle to Montreal and Beyond: The Battle Royale of the 21st Century," a paper from The Institute for Agriculture and Trade Policy, Minneapolis, February 2000.

4. Ann Veneman, USDA statement, August 30, 2002.

5. Hope Shand, Pew Initiative on Food and Biotechnology, Spotlight, Of Famine and Food Aid: GM Food Internationally, October 2002.

6. "Poverty and Transgenic Crops," *Nature,* August 8, 2002, 569.

7. Simon Robinson, "To Eat or Not to Eat: As Zambia Starves and the E.U. Battle over Genetically Modified Food Aid in Africa," *Time* (European edition), December 2, 2002.

8. Ibid., 7.

9. Kurt Kleiner, "Fields of Gold: Biotech Cash Benefits May Not Be What They Seem," *New Scientist,* June 22, 2002: 11. Leonard Gianessi, National Center for Food and Agricultural Policy, partly funded by Monsanto and the Biotechnology Industry Organization.

10. Ibid. Also Charles Benbrook, The Bt Premium Price: What Does It Buy?

The Impact of Extra Bt Corn Seed Costs on Farmer Earnings and Corporate Finances, Paper, Benbrook Consulting Services, Sandpoint, Idaho, February 2002.

11. Andy Coghlan, "Weeds Do Well out of Modified Crops," *New Scientist,* August 17, 2002:11.

12. Rebecca Goldburg, quoted by Bob Holmes, "Dangerous Liaisons," *New Scientist,* August 31, 2002. 38–41.

13. Andrew Pollack, "Widely Used Crop Herbicide Is Losing Weed Resistance," *New York Times,* January 14, 2003: C1.

14. Proceedings of the National Academy of Sciences, www.PNAS online, DOI:10.1073/pnas.252637799; and BBC News online. http://news.bbc.co.uk/2/hi/science/nature/251295.stm.

15. Andy Coghlan, "Sweet Genes Help Rice in a Drought," *New Scientist,* November 30, 2002, 10.

16. BBC News Online, "GM Tomato to Fight Disease," May 2, 2001. Research led by Martin Verhoyen, Unilever Research, Sharnbrook, U.K. http://www.biotech-info.net/GM_tomato.html.

17. Joel Cohen and Robert Paarlberg, "Explaining Restricted Approval and Availability of GM Crops in Developing Countries," *AgBiotechNet,* October 2002, 6.

18. Ibid., 6. Monsanto had renounced use of the technology after intervention by Gordon Conway, president of the Rockefeller Foundation, but the fear persisted.

19. Philip Cohen, "Begone Evil Genes," *New Scientist,* July 6, 2002, 33–36. The Exorcist was the brainchild of Pim Stemmer and his colleague Robert Keenan, of the biotech company Maxygen of Redwood City, California.

20. Sue Mayer of Greenwatch U.K. in Philip Cohen, "Begone Evil Genes," *New Scientist,* July 6, 2002: 33.

21. Anne Simon Moffat, "Can Genetically Modified Crops Grow Greener?" *Science,* October 13, 2000, 253.

22. Author interviews, April 2001; June 2001. See also Ehsan Masood, Opinion interview, *New Scientist,* October 21, 2000; Elizabeth Finkel, "Australian Center Develops Tools for Developing World," *Science,* September 1999; Barnaby J. Feder, "New Method of Altering Plants Is Aimed at Sidestepping Critics," *New York Times,* Science Times, February 29, 2000, 1.

23. John Doebly, *Nature,* March 18, 1999, 236.

24. Scot Kilman and Roger Thurow, "Africa Could Grow Enough Food Itself: Should It?" *Wall Street Journal,* December 3, 2002, 1 and 14.

25. Andy Coghlan et al., "Beyond Organics," *New Scientist,* May 18, 2002, 33–47.
26. Glenn Garelick, "Taking the Bite out of Potato Blight," *Science,* November 29, 2002; 1702–4.
27. Yves Savidan, author interview, April 25, 2001. See also Charles Benbrook, "Who Controls and Who Will Benefit from Plant Genomics?"; paper to AAAS annual meeting, Washington, D.C., February 19, 2000.
28. Justin Gillis, "Cultivating a New Image, Firms Give Away Data, Patent Rights on Crops," *Washington Post,* May 23, 2002.
29. Jeffrey Burkhardt, "Biotechnology's Future Benefits: Prediction or Promise?" *AgBioForum* 5(2): 20–24; Peter Raven, "Presidential Address, AAAS," *Science,* August 9, 2002; Susan McCouch, author interview, October 29, 2002; Rebecca Goldburg, author interview, 2002; Calestous Juma, "How Not to Save the World," *New Scientist,* September 28, 2002, 24; Anthony Trewavas, "Malthus Foiled and Foiled Again," *Nature,* August 8, 2002, 668–70; David Tilman et al., "Agricultural Sustainability and Intensive Production Practices," *Nature,* August 8, 2002, 671–77; R. Hails, "Assessing the Risks Associated with New Agricultural Practices," *Nature,* August 8, 2002, 685–88.
30. Klaus Leisinger, Karin Schmidt, and Rajul Pandya-Lorch, "Six Billion and Counting. Population and Food Security in the 21st Century," International Food Policy Research Institute, 2002, 8–9, and Per Pinstrup-Anderson and Ebbe Schiøler, *Seeds of Contention,* 68–71.
31. Vandana Shiva, "The Seed and the Spinning Wheel: The UNDP as Biotech Salesman," Reflections on the Human Development Reprint, July 25, 2001.

ACKNOWLEDGMENTS

I owe my original interest in genetically modified foods to two friends of many years. Susan Dooley, a writer and small farmer in Maine, was worried that genetic engineering would eventually cause the loss of her traditional seeds. Susan Sechler, as head of the Global Inclusion Program at the Rockefeller Foundation in New York, was concerned about the increasing control large corporations had over biotech and agriculture, and the effect on poor farmers in the developing world. She urged me to apply for a Rockefeller Foundation writing and research grant, which I did. The grant gave me the freedom to explore biotech agriculture.

In those early stages, I was especially fortunate in finding two tutors—Richard Jefferson and Susan McCouch—who gave me an expert introduction to this world and kindly shared their thoughts on the international implications of biotechnology. The Swiss botanist Klaus Ammann passed along his special insights into the role of the corporations. Dan Charles, an old friend who wrote an excellent book about Monsanto's part in this drama, *Lords of the Harvest,* was extremely generous with his knowledge and his contacts. Many other experts in biotech agriculture were also generous in giving up their time to talk to me. I thank particularly Ingo Potrykus, Peter Beyer, Roger Bull, Swapan Datta, Bob Goodman, Rosie Hails, Gurdev Khush, Terri Lomax, Alan McHughen, Margaret Mellon, Wayne Carlson, Kristin Dawkins, Rebecca Goldburg, Major Goodman, Yves Savidan, Charles Spillane, Larry Stenberg, Hope Shand, Gary Toenniessen—the oracle of progressive agriculture—and Brian Wright. Several people kindly invited me to eavesdrop on their agbiotech seminars and conferences. My thanks for such invitations to Marianne Ginsburg of the German Marshall Fund; Dave Ervin, Terri Lomax, and Michael Rodemeyer; to Mary Kaldor and Diane Osgood; Lucia Colombo; David Hastings and Calestous Juma; the John Innes Center in the U.K. and the Food and Drug Law Institute in Washington, D.C. Others provided valuable grounding in these complex issues, including Andreas Bieberbach, Mauricio Bellon, Julien Berthaud, John Dodds, Ignacio Chapela, Luis Herrera Estrella, Laurie Friedman, Paul Heisey, Katerina Jenny, Wayne Hanna, Gurdev Khush, Anatole Krattiger, John Losey, Mwananyanda Lewanika, Bernard Marantelli, Tamara Naumova, David Poland, Gwyn Prins, Yves Savidan, Victor Sokolov, and Florence Wambugu. As always, Natalia Alexandrova was invaluable in Moscow, where Yuri Vavilov so generously shared his memories and his father's papers. In London, my friend Zhores Medvedev let me rummage through his Russian library.

The Meridian Institute's Agriculture Biotechnology Intranet was an invaluable independent Web source, as was the Pew Initiative on Food and Biotechnology and Ag BioTech InfoNet. I thank CAB International for providing review copies of their publications, Genetic Resources Action International (GRAIN) for research papers, the ETC Group (formerly RAFI) for their research and Pat Mooney for his provocative headlines, and the Union of Concerned Scientists. My thanks to Susan Dooley, Bob Goodman, and Richard Jefferson for reading the manuscript at various stages and for their valuable comments.

For various other kinds of help and advice, I thank Ben Bradlee, Dan Conable, Dick Daynard, Nicholas von Hoffman, Mary Ellen Barton, Philip Jacobson, Pedro Meyer, Fabio di Paolo, and John Pringle. The production team at my publishers—Anja Schmidt, Jonathan Jao and my copy editors, Peg Haller and Nancy Inglis—were superb. Alice Mayhew, my friend and editor, was a constant source of enthusiasm, wit, and wisdom. The love and counsel of my wife, Eleanor Randolph, sustained me through the writing of this book, as before, and our daughter Victoria, once again, tolerated with great equanimity the general disruption of family life.

INDEX

ABOUT THE AUTHOR

Peter Pringle is the author and coauthor of several books, including the best-selling *Those Are Real Bullets: Bloody Sunday, Derry, 1972*. He has written for *The New York Times, The Washington Post, Atlantic, New Republic,* and *Nation*. He lives in New York City with his wife, Eleanor Randolph, a *New York Times* editorial writer, and their daughter.